EDUCATIONAL OPP

International Population Studies

Series Editor: Professor Philip Rees,
School of Geography, University of Leeds, UK

This book series provides an outlet for integrated and in-depth coverage of innovative research on population themes and techniques. International in scope, the books in the series will cover topics such as migration and mobility, advanced population projection techniques, microsimulation modeling, life course analysis, demographic estimation methods and relationship statistics.

The series will include research monographs, edited collections, advanced level textbooks and reference works on both methods and substantive topics. Key to the series is the presentation of knowledge founded on social science analysis of hard demographic facts based on censuses, surveys, vital and migration statistics. All books in the series are subject to review.

Educational Opportunity
The Geography of Access to Higher Education

ALEXANDER D. SINGLETON
University College London, UK

Routledge
Taylor & Francis Group

LONDON AND NEW YORK

First published 2010 by Ashgate Publishing

2 Park Square, Milton Park, Abingdon, Oxon OX14 4RN
711 Third Avenue, New York, NY 10017, USA

Routledge is an imprint of the Taylor & Francis Group, an informa business

First issued in paperback 2016

British Library Cataloguing in Publication Data
Singleton, Alexander D.
  Educational opportunity : the geography of access to higher
  education. -- (International population studies)
  1. Universities and colleges--Public relations.
  2. Universities and colleges--Admission. 3. Discrimination
  in higher education. 4. Educational equalization.
  5. Education--Regional disparities.
  I. Title II. Series
  659.2'9378-dc22

Library of Congress Cataloging-in-Publication Data
Singleton, Alexander D.
  Educational opportunity : the geography of access to higher education / by
Alexander D. Singleton.
    p. cm. -- (International population studies)
  Includes index.
  ISBN 978-0-7546-7867-0 (hardback)
1. Education, Higher--Social aspects--Great Britain. 2. Education--Demographic
aspects--Great Britain. 3. Educational equalization--Great Britain. I. Title.

  LC191.98.G7S55 2010
  378.1'61--dc22

                                                                        2010005994
  ISBN 978-0-7546-7867-0 (hbk)
  ISBN 978-1-138-27239-2 (pbk)

Contents

List of Figures

List of Tables

For Melissa

Chapter 1
Access to Higher Education

Higher Educatiaon participation rates have long been known to differ between societal groups. However, despite extensive research examining the underlying causes of these inequalities, and significant government funding designed to create a more egalitarian system, a recent UK National Audit Office report highlights that some groups still remain significantly under-represented (Comptroller and Auditor General 2008). Additionally, although there has been much research into the binary decision of attending Higher Education, there is much less research into the more complex set of decisions which lead to this choice. For example, what are the differentials in course or institution participation rates between different groups in society? How can we understand and model these complexities? As such, the overarching aim of this book is to investigate a better way of representing the multiple social, spatial and temporal processes which shape access to Higher Education.

In organisational terms, the agendas of widening participation, extending access and institutional marketing present common challenges. They aim to devise better ways of reaching potential students who are appropriately qualified and motivated to pursue and successfully complete their chosen course offerings. However, few Higher Education institutions have communication strategies that are tailored towards reaching a full range of potential students who could benefit from their range of existing subject and course offerings, and no ways of making innovation in course design responsive to potential student demand. University and college marketing initiatives are often unsystematic in the ways in which they target schools and colleges for widening participation, and uncoordinated in the presentation of their full institutional profiles of subjects of study in these activities. This book addresses the changing Higher Education policy-setting arena and presents a systematic framework for widening participation and extending access in an era of variable fees. It illustrates how Higher Education data and publicly available sources might be combined to enable institutions to move from piecemeal analysis of their intakes to institution-wide strategic and geographically linked market area analysis for existing and envisaged subject and course offerings.

The book is divided into nine chapters which explore the following topics:

- The geographies of access to Higher Education.
- Educational data sources, tools and profiling techniques.
- The use of geodemographics in Higher Education.
- Supporting widening participation initiatives in Higher Education.
- Access trends in Higher Education: Are things getting better?
- The role of schools.

The development of contemporary Higher Education systems is described through analysis of historical changes in participation and access rates. The primary focus is upon the UK experience, but many of the issues of widening participation are salient in other systems. Moreover, the techniques that are described are generic to most all developed economies. Although the UK Higher Education system has been transformed into one which serves the mass market, it will be shown how there are still persistent under represented groups. Within this context, the methods of current data collection and their organisational structures are described to highlight complexities, overlaps, and demonstrate where problems of coordination can cause duplicate or missing data. It is shown how, despite growing demands and requirements for information created by integrating these disparate datasets, these are not currently being provided through centralised services to key stakeholders in Higher Education. Geodemographic methods are introduced as a method of organising such information, and it is suggested how concepts of social capital can map into these quantitative variables that permit generalisations. Using this information base and analytical toolkit, it is demonstrated how the stakeholders in Higher Education may currently be making uninformed policy-sensitive decisions because of a lack of appropriate decision support tools. The discussion begins with an overview of the spatial and social complexities of Higher Education, including detailed description of a number of variables, such as distance travelled to accept a place, the type of institution attended, attainment differentials and course choice. Later in the book, the concept of classification is extended to examine how more relevant geodemographic indicators for Higher Education can be built and evaluated. This analysis challenges the implied assumption held by commercial classification builders that an individual's use of public services is analogous to the ways in which consumers use private goods. The final part of the book examines whether Higher Education inequalities are getting better or worse and how these may be addressed through engaging with stakeholders in Higher Education.

Chapter 2
A Meritocratic Marketplace?

Historical Development of Access and Participation Inequalities

Inequality in participation and access to Higher Education has a long history extending back to mediaeval times. The interplay between external influences such as industrialisation and centralised policy change has shaped spatial and societal Higher Education access patterns. To understand the heterogeneous nature of contemporary Higher Education, it is important that these historical patterns of access be identified.

Towards the end of the 12th century, the English Universities of Oxford and Cambridge were established. Cobban (1999) states that before the 16th century those students attending these institutions were of middle to lower social condition, and were structured less on wealth or class, and markedly removed from the rigidity of the social structures outside academia. However, the system became more elitist towards the end of the 15th century, with an influx of students from noble births. This contrasted with the acceptance patterns of European institutions which, that throughout the mediaeval period, had a considerable number of noble scholars. Participation was limited during the mediaeval period with only 3% of the male population attending university (Ainley, 1994).

The Civic Universities were established during the 19th century as a solution to the growing educational demands of rapidly developing industrial cities. These institutions developed a very different pedagogy to the mediaeval institutions, with a curriculum centred on meeting the needs of a newly growing industrial society. Table 2.1 plots the establishment of the Civic Universities.

Most of these first phase Edwardian "redbrick" institutions originated from older institutions such as working men's colleges or institutions (Ross, 2003). However, the second wave of institutions established close to World War I had only some of their origins in these institutions. Originally these institutions would teach University of London Degrees, but gradually were granted independent status (Ross, 2003). These institutions included Reading, Hull, Nottingham, Southampton, Exeter and Leicester.

Anderson (1992) describes the class structure in the Civic University system as following three distinctive phases of development. During the early phase the institutions continued the traditional task of serving the older land owning professional elite. However, during the mid phase around 1860, these entry profiles began to change, with institutions adapting to the needs of the industrial society with an increasing class mix. The late phase proceeded with a shift towards a more diverse and middle class professional system, leading to a faster increase in student

Table 2.1   The development of the first phase Civic University

University	Date	Notes
Durham University	1833	
London University	1836	Amalgamation of King's College and University College which were established 10 years earlier.
Victoria University	1880	Constituted of colleges in Manchester, Leeds and Liverpool. These split in the early 1900s.
Birmingham	1900	
Sheffield	1905	
Bristol	1909	

Source: Adapted from Ross, 2003.

numbers, and a greater expansion as middle class occupations grew. The changes occurring in the civic system challenged the traditional universities of Oxford and Cambridge, therefore forcing them to reform between 1850 and 1870.

After World War II, there were various Higher Education reforms which aimed at treating all scholars equally, achieved through mandatory grants, and a selection process for university based upon the Advanced Level and Scottish Higher examinations.

UCCA (1994) discusses how two major concerns dominated university admission staff during the 1950s. Firstly, applications for undergraduate courses were tending to increase year by year as more young people were entering sixth forms. A second concern was demographic, with the expectation that the university population was about to increase rapidly as the post-war baby boom reached Higher Education participation age. UCCA (1994) quantifies this growth in applications between 1955 and 1960, with Nottingham University reporting an increase of 150% and Leeds University 130%. This trend was boosted by individuals applying to more than a single university, which occurred to a lesser extent before World War II. Ainley (1994) observes that, as a result, Higher Education had grown to 7.2% of the age cohort by 1962.

In 1961 the Universities Central Council on Admission (UCCA) was set up as the centralised admissions agency for UK courses of full time Higher Education. The key aims were to create a fairer and more objective system of entrance that would also reduce the administrative burden on individual institutions. It should be noted that when Polytechnic Colleges were established during the 1980s, the Polytechnic Central Admissions Service (PCAS) rather than UCCA managed their applications. Figure 2.1 shows how during the period 1962 to 1990 there was a steady increase in both home and overseas applications and acceptances.

In 1963 a report compiled under Lord Robbins (Robbins, 1963) set out recommendations that courses of Higher Education should be made available to all of those who were qualified by attainment to pursue them and who wished to do so. It established that the majority of this Higher Education would be based in universities, although it also observed that there was also a growing need for

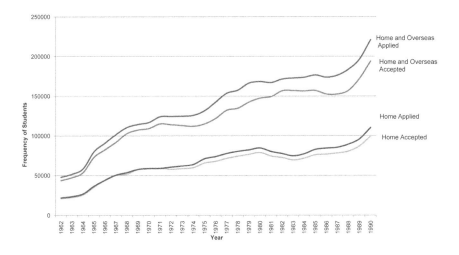

Figure 2.1   Applications and acceptances 1962–1990

Source: UCAS.

provision of more vocational education. In 1966 a White Paper entitled "A Plan for Polytechnics and Other Colleges" was delivered in Parliament (Secretary of State for Education and Science, 1966). In this report it was suggested that:

> The Government have committed themselves to an even greater expansion of Higher Education than was forecast in the Robbins report, and in this document announce their intention of developing a new sector of Higher Education within the Further Education system, to complement the universities and colleges of education…The generic term for these new centres is to be "polytechnics". (Secretary of State for Education and Science, 1966:2)

These recommendations established a binary divide between traditional universities and the new type of institution called polytechnic colleges (Ainley, 1994). Polytechnic colleges were established to push comprehensive principles into Higher Education by offering a broader range of opportunity to a wider base of students, feeding local community needs through local education authority control, and eroding "the elitist character of British Higher Education" (Carr, 1998:275). The rationale for polytechnic education can be traced back to the writings of Marx and Engels, where real-life problems are tackled and reflected on jointly by teachers and the taught. Ainley (1994) discusses that this "universities for all" principle pioneered what the Polytechnics' idealistic founders referred to as liberal vocationalism, and that this shift in pedagogy reflected changes that had occurred in contemporary comprehensive and community schools. The polytechnic education aimed to make available opportunities to qualify for occupations on equal terms with those educated selectively. Thus, the polytechnic system began to widen access with the percentage of full and part time students growing to 12% of the age range by the end of the 1970s (Ainley, 1994).

During this period, there was a significant shift in policy towards reshaping Higher Education under conditions of severe resource restraint, beginning with publication of the 1981 Public Expenditure White Paper (Eurydice, 2004). This stated:

> This is likely to oblige institutions to review the range and nature of their contribution to Higher Education. It is also likely to lead to some reduction in the number of students admitted to Higher Education with increased competition for places.

This resulted in a reduction in public sector expenditure in Higher Education of around 8% over three years. The rationalisation of Higher Education continued until around 1987 when the publication of the White Paper "Higher Education: Meeting the Challenge" set out policy changes to increase participation rates and widen access to Higher Education for non traditional applicants, such as mature students, or those without A-Levels (Eurydice, 2004).

The binary divide between the two systems had become blurred, with Polytechnics moving towards the academic norms of universities, while universities were becoming less elitist and inward looking. Jenkins (1995) argued that "polytechnics had not become universities, universities had become polytechnics".

Another White Paper was released in 1991 entitled "Higher Education, A New Framework" (Department for Education, 1991), which set out proposals to remove the binary divide between universities and polytechnic colleges, and was passed through Parliament as part of the 1992 Higher Education Act. Carr (1998) describes the three key aims of this post-binary policy as:

- Creating funding agencies that would be more sensitive to the secretaries of state rather than acting as buffer between academics and the state.
- Continuing moves towards a competitive funding system, where former polytechnic colleges could compete for a share of £680m allocated to universities for research.
- Create a dual system of quality assurance incorporating the old universities, and in time, generating league tables to underpin student choice in the new unified Higher Education system.

This influx of institutions and students into the Higher Education system, in a decade of underfunding, has led to large deficits in the finances of many of the UK's leading institutions. Funding has been improving in recent years, although in 2002/03 there were still 48 institutions in deficit (MacLeod, 2004). Institutions created numerous coping strategies to deal with these debts such as increasing the number of student places. The fall in public funding per student is shown in Figure 2.2, first published in Dearing (1997).

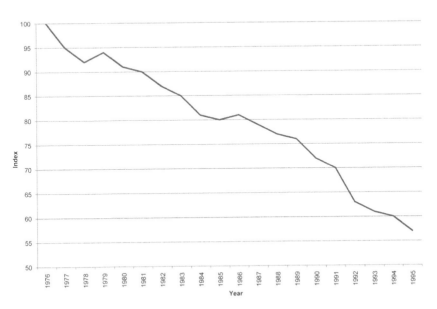

Figure 2.2   **Index of public funding per student in Higher Education 1976–1995**

Source: Dearing, 1997.

In line with the post-binary policy changes, PCAS and UCCA were amalgamated during 1993. 1994 entry was managed by a new organisation, the Universities and Colleges Admissions Service (UCAS). The effect of this amalgamation can be seen in the apparent massive upturn in both applications and acceptances during the 1994 recording period (see Figure 2.3). There is a second increase around 1997 where the Art and Design Admissions Registry (ADAR) was amalgamated with UCAS. Post-1998 there is a continual and steady growth in the frequency of both applications and acceptances.

When considering participation, it is important to understand how the addition of institutions to what is being classified as Higher Education can affect these rates. Figure 2.4 below shows how the number of UCCA and UCAS institutions has changed dramatically since 1963.

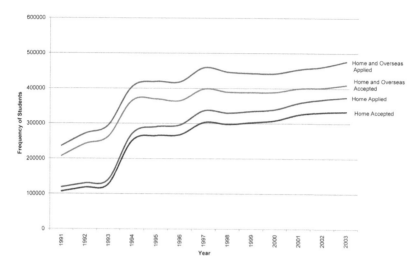

Figure 2.3   Applications and acceptances 1991–2003

Source: UCAS.

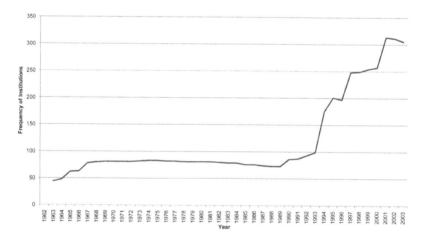

Figure 2.4   UCCA and UCAS institutions 1962–2003

Source: UCAS.

The Dearing Report (1997) investigated how the purpose, shape, structure and funding of Higher Education should meet the needs of the United Kingdom for the following 20 years. This report was far reaching in its objectives and included recommendations about:

- Demand for Higher Education
- Widening Participation
- Students and Learning
- Content of Programmes
- Qualifications and Standards
- Research and Funding
- Regional and Local Impact of Institutions
- Human Resources Issues
- Funding

In 1997, the newly elected Labour Government acted on the Dearing Report with recommendations to scrap the remaining grants and replace them with student loans linked to parental income. There was also the announcement that means tested tuition fees would be introduced to the sum of £1,000.

The Dearing Report highlighted issues of access rather than participation, discussing that there remain groups in the population who are underrepresented in Higher Education, including lower socio-economic groups and certain ethnic minorities. "Participation" and "access" are subtly different terms and in public policy discussion they are often used interchangeably. However, in this book, "participation" will refer to increasing the absolute frequency of students attending Higher Education, whereas "access" refers to readdressing the balance between the constituent groups making up the total participation.

When the Labour Party was re-elected in 2001, an ambitious 50% Higher Education participation target was set. Although this target was described as arbitrary (May, 2003), figures suggest current participation rates may have reached 45% of the relevant age cohort (*Economist*, 2004). However, the extent to which this can be apportioned between demographic change, improving A-Level performance and widening participation initiatives, is debatable.

Many institutions now face financial crisis because funding has not kept pace with increasing student numbers. Funding in Higher Education has fallen in real terms by 38% since 1989, resulting in massive underinvestment (Brown and Piatt, 2001). The Oxford Centre for Higher Education Policy Studies (OxCHEPS) estimates the costs to educate a 2003 student at Oxford University to be around £18,600 per annum, where only 6% of this cost comes from student fees. Of the remaining 94%, around half comes from private sources such as endowments, with the government contributing the other half through the Higher Education Funding Council for England (HEFCE) (Palfreyman, 2004).

The re-election of the Labour Party in 2001 came with a manifesto promise stating "we will not introduce 'top-up' fees and have legislated to prevent them" (Blair, 2000). In 2003 the White Paper "The Future of Higher Education" was published. This paper had two interrelated purposes: to outline the funding gap required to maintain international teaching and research standards; and also to create conditions for equality of access. A bill containing the recommendations from this paper was only narrowly passed (by five votes) through Parliament in 2004. The changes included:

- A capped variable tuition fee up to the value of £3,000
- Establishing an access regulator
- Introducing maintenance grants of between £1,000–£1,500 for poorest 30%
- Poorest students receive fee remission for the first £1,500 (£1,200 by state and £300 by institution)
- Student loans to be aligned with the real cost of living
- All student debt dropped after 25 years

The previous discussion has shown that increasing participation has almost certainly occurred; however, the extent to which it has "widened" is debateable (Farr, 2002). UCAS statistics show that from 1996 to 2002, home applicants for full time degrees rose by 26% (UCAS, 2003a). However, if the acceptances to both UCCA and UCAS are examined by the National Statistics Socio Economic Classification (NS-SEC) and previous equivalents such as Occupational Group and Social Class, it is possible to see that the access patterns between groups have been relatively static throughout the last 40 years. However, it should be noted that temporal analysis of UCCA/UCAS data should be used with the caveat that the institutions from which the total population is derived do not remain static, with some institutions entering, leaving or amalgamating each year. The effect of these changes on the total intake of institutions was show earlier in Figure 2.4.

During the period 1968–1978 Occupational Group (Rose, 1995) was used to classify accepted students' parental occupation. The pattern is relatively static, perhaps with a slight increase of Professional and Technical occupations and a decrease in Manual (see Figure 2.5).

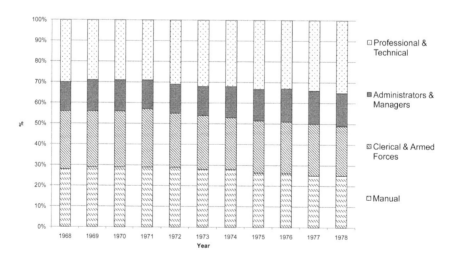

Figure 2.5   Occupational Group 1968–1978

Source: UCAS.

Figure 2.6 shows the change in the Social Class (Rose, 1995) of accepted students between 1980 and 2001. The 1980–1993 data were extracted from UCCA statistical bulletins and were based upon direct questioning of a random sample of students. These data therefore do not record students with parental occupation classified as unknown. In the UCAS acceptances during the period 1994–2001, these data relate to all students admitted through the UCAS process (not a sample), and as such, this accounts for the sudden change in profile between 1993 and 1994. In the 1980–1993 period the access rates remained relatively static with professional and intermediate social classes dominating acceptances. During the period 1993–2001, once the Polytechnic and University admissions systems had been combined, a slightly larger percentage of total acceptances was derived from the lower social class groups. Also during this period there was a growing number of unknown classifications, perhaps as a factor of the Social Class schema becoming progressively out dated.

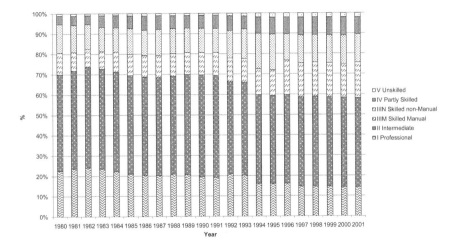

Figure 2.6   Social Class 1980–2001

Source: UCAS.

In 2002 a new Socio-Economic Group classification[1] was adopted. From looking at these proportions, it appears that, despite government intervention, the socio-economic profile of acceptances remains reasonably static, perhaps with some growth in the frequency of students classified as "Unknown" (see Figure 2.7). This theme is investigated later in the book.

_____

1   www.statistics.gov.uk/methods_quality/ns_sec/cat_subcat_class.asp.

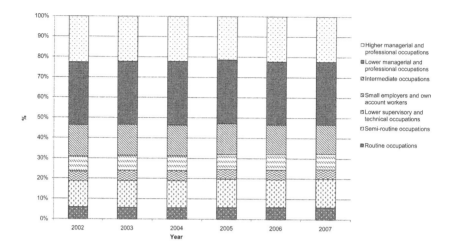

Figure 2.7   Social-Economic Group 2002–2007

Data Representation and Reduction

In the previous section data have been used from a variety of sources to illustrate a picture of both the aggregate growth in Higher Education sector participation and how these patterns are stratified by measures of social class. The sources of data used for these analyses were derived at an aggregated scale, through pre-printed publication or online analytical tools. However, unit postcode level data for both Higher Education and schools have been made available for the analysis conducted throughout this book. These data are voluminous and have national coverage across a range of levels.

The UK education system offers a number of routes and potential outcomes in terms of qualification type, progression routes and organising bodies. The system is divided into a series of levels from Primary through to Postgraduate Degree and at present is compulsory until the age of 16. Assembly and interpretation of educational statistics is difficult because of the plethora of definitions of educational qualifications, the overlapping responsibilities of data collecting organisations, and the inability to track the educational careers of individuals through unique identifiers across all datasets. The analysis presented in this book uses various national Higher Education and FE datasets that are collected by a number of different agencies. These are summarised in Table 2.2.

Table 2.2   Educational data collection agencies

Sector	Data Sets	Agency	Contents
FE	Individualised Learner Record (ILR)	Learning and Skills Council (LSC)	Data relating to English post-16 education and training
FE/HE	Courses, Applicants and Acceptances Database	Universities and Colleges Admissions Service (UCAS)	UK and international dataset holding information on UCAS applications, offers and their outcomes
School	Pupil Level Annual School Census (PLASC)	Department for Education and Skills (DfES)	UK dataset relating to individual students in all schools
HE	Student; Student First Destinations; Staff; Finance and the Non-credit-bearing Course Records	Higher Education Statistics Agency (HESA)	Various UK datasets covering various aspects of staff and students in Higher Education

The Higher Education Data Trail

Higher Education data collection and dissemination are not managed by a single organisation. These data are predominantly created through the applications process to full time Higher Education and involve a range of transactions between individuals and national organisations. A simplification of how the UCAS admissions process creates a range of data is shown in Figure 2.8.

The Universities and Colleges Admissions Service (UCAS) centrally manage the application process for all full time courses of Higher Education in the UK. The UCAS nomenclature defines "Applicants" as those individuals seeking entry to Higher Education through UCAS. Applicants to Higher Education make an initial selection of six choices (applications) which each consist of an institution, course and campus selection. UCAS collect these data alongside various other attributes of the individual applicant, e.g. age, gender, address etc. The majority of applicants now submit their applications electronically and much of the data processing is automated. For the raw UCAS database to remain accurate, this requires applicants to have both submitted data truthfully and with minimal error. Where errors are detected though automated checking, these are flagged for human intervention, and data processing operatives correct these details where possible. The main application cycle begins in October of each year and runs through to a deadline at the end of June in the following year. During this period, applicants receive offers or rejections from their six or fewer applications. These decisions are communicated on behalf of the institutions through UCAS, as conditional offers, unconditional offers or rejections.

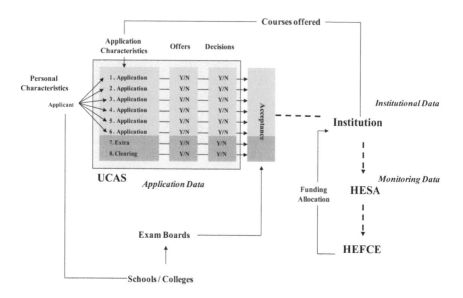

Figure 2.8   A simplified model of Higher Education data flows

Conditional offers usually specify a particular grade attainment or subject requirement. Unconditional offers are those with no conditions attached, and are usually made where an applicant already has prior qualifications, or where candidates have been judged to have exceptional promise or aptitude. Once institutions have responded to all applications from an applicant, the applicant can reply, either accepting or rejecting the offers received, again communicating these decisions through UCAS. For those students who receive no offers before the main scheme closing date, a further option called UCAS Extra is made available. This allows applicants to approach institutions advertising courses within the Extra scheme and make repeated applications until they successfully receive an offer. From the main scheme an applicant can hold a maximum of two offers into the summer confirmation phase in mid August when exam results are delivered to students and also communicated to UCAS. This allows those applicants with open conditional offers to be automatically confirmed or declined, based on a comparison between attainment and offer requirements. Those applicants who apply after the main application scheme closing date, those who do not receive offers from any institutions, and those who do not meet their offer requirements are placed into a process called clearing, where institutions advertise their remaining course places. Negotiation then occurs between applicants and institutions directly, and if places are found, these are again communicated through UCAS. It is through this yearly application cycle that the majority of Higher Education data are collected for the UK. Institutions are supplied electronically with data collected by UCAS for their admitted applicants, and these in turn populate internal acceptance databases or student records. Once these data are within the institutions, they can be updated or amended as necessary, although such subsequent changes are not returned to UCAS. For undergraduate admissions, UCAS only collect data on full time admissions, and only from those institutions within the UCAS scheme. This currently excludes the University of Buckingham, Birkbeck College (University of London) and the Open University. The UCAS scheme also includes a wide range of smaller colleges offering Higher Education level full time qualifications, often referred to as university sector colleges.

Institutions are funded through a mixture of public funding through central government, private enterprise or investment, entrepreneurial activity and interest from endowments. Following UK devolution in September 1997, centrally managed funds are distributed through separate organisations in England, Scotland, Wales and Northern Ireland. These are shown in Table 2.3.

Table 2.3   Higher Education funding councils

Country	Funding Agency
England	Higher Education Funding Council for England (HEFCE)
Scotland	Scottish Higher Education Funding Council (SFC)
Wales	Higher Education Funding Council for Wales (HEFW)
Northern Ireland	Northern Ireland Executive: Department for Higher and Further Education, Training and Employment (DFHETE)

In order to apportion funds appropriately, Higher Education funding councils require data on the size, shape and performance of institutions. These data are acquired through the Higher Education Statistics Agency (HESA), which is the "central source for collection and dissemination of statistics about publicly funded UK Higher Education" (HESA, 2006). Each Higher Education institution in the UK that receives public funding is required to submit an annual "HESA Return", which are datasets of a standard format detailing those students within the institution. Various details are collected: however, the majority of the undergraduate student record is derived from the UCAS data supplied at the end of the application cycle. Institutions are encouraged to maintain and update these data as they can have bearing on those funds available for the following academic year. For example, the calculation for widening participation funding for young participants (aged <21) is based on students being grouped into participation rate bands aggregated into the ward boundary in which they live (HEFCE, 2005). Therefore, if an institution has erroneous or missing postcode data in its applicant records, funding may be misappropriated or lost because of geocoding errors when converting these postcodes into spatial locations. Therefore, the two key sets of data which exist for the Higher Education sector are the UCAS and the HESA datasets, the former of which is specifically associated with undergraduate admissions. HESA and UCAS have made data available for this study and the variables that are of interest will be discussed later in the chapter.

What is Higher Education?

The previous section discusses how data on Higher Education are created and flows between organising bodies and stakeholders. However, before further discussion on other educational data sources it is essential to define what is meant by "Higher Education" and the types of institution that are included in Higher Education datasets. Higher Education is not necessarily university education and the title an institution holds does not determine its classification as either Higher or Further Education (now the Learning and Skills Sector) because the use of the term "University" is legislated differently depending on the age of an institution. Older (pre-1992) Higher Education Institutions operate under a Royal Charter, whereas newer (post-1992) ones operate under an Instrument of Government and Articles of Government. Both are now managed by a part of government called the Privy Council who are responsible for changes to institutions' constitutions, and also the use of the name "University" and "University College" within their title.

As discussed earlier, UCAS processes UK and overseas applications for Higher National Diplomas, Degree and Foundation Degree qualifications for its member institutions. These institutions could be broadly classified as Higher Education, with a number of exclusions including the Open University and Birkbeck College, both of whom operate distance learning and part time courses, and the University of Buckingham, as this institution is privately funded. Higher Education institutions, as defined by UCAS can be "University" or "University Sector Colleges": however there is no standard definition of either group. University Sector Colleges are often not universities by title, but do run degree or higher-level qualification in conjunction with a university. Conversely, there are also universities that offer FE qualifications alongside degrees. This confusion is further highlighted in the National Qualification Framework (see Table 2.4). In this framework, qualifications of similar attainment are grouped into a series of six levels, with level three the normal progressive step into Higher Education. Therefore, a further definition of a Higher Education institution could be one that supplies level four courses or above. However, this is also problematic, as many Further Education colleges offer vocational qualifications up to levels four or five, but may not offer general qualifications beyond level three.

UCAS mainly manages level four courses (HND and Degree), and as such, many courses at University College Sector institutions are not administered through the UCAS system, with these courses being classified as Further rather than Higher Education.

Table 2.4   The National Qualifications Framework

Levels	General Qualifications	Vocation Related Qualifications	Occupational Qualifications
Higher Level 5	Higher Level Qualifications		NVQ Level 5
Higher Level 4	Degree	Foundation Degree, HND	NVQ Level 4
Advanced Level 3	GCE A-Level/AS	Vocational A-Level (Advanced GNVQ)	NVQ Level 3
Intermediate Level 2	GCSE Grades A*–C	Intermediate GNVQ	NVQ Level 2
Foundation Level 1	GCSE Grades D–G	Foundation GNVQ	NVQ Level 1
Entry Level	Certificate of Educational Achievement		

Source: UCAS.

Classifying Institutional Groups

In the previous two sections it was argued that there can be duplication between sector datasets and that the definition of specific education sectors can be convoluted; specifically given the historical context of Higher Education sector growth as measured by admissions through UCAS and its predecessors. Two key events, the amalgamation of Universities and Colleges Central Admissions (UCCA) and Polytechnic Colleges Admissions Service (PCAS), and then UCAS and Arts and Design Admissions Registry (ADAR) resulted in the analytical agglomeration of large constituent groups of very distinctive institutions, the Universities, the Polytechnics and the Arts Colleges. However, between these main divisions there are a diverse set of policy initiatives which have resulted in the establishment of a finer taxonomy of institutions. Through a joint project between the University of Oxford and UCAS, the taxonomy in Table 2.5 was created.

Although these groupings provide a useful framework to understand the history of how different types of university were established, they are not agglomerations used in policy making, nor lobbying. Groupings of institutions which have formed outside of historical coincidence, and with a specific remit include:

- The Russell Group
- The 1994 Group
- N8 Group

The Russell Group[2] is the largest aggregation of research intensive institutions, set up in 1994 as a lobbying group to promote their interests to government and associated bodies. This group accounts for 65% of the total UK university research

---

2   www.russellgroup.ac.uk.

Table 2.5   The Oxford University classification of institutions

Category	Some Examples
Ancient Universities	The University of Cambridge, The University of Oxford.
Old Universities (up to c. 1900)	University of Durham, The University of Birmingham, Cardiff University.
Old Universities c. post-1900)	The University of Leicester, The University of Hull, The University of Southampton.
Robbins "New Universities"	The University of York, The University of Essex, The University of Sussex.
Robbins "Technological Universities"	Loughborough University, The University of Salford, Brunel University, The University of Bath.
Post-1992 Universities (ex Polytechnic)	The University of Plymouth, Oxford Brooks University, Middlesex University.
Post-1992 Universities (ex HE College)	Cranfield University.
HE Colleges and University Colleges – Generalist	Bath Spa University College, Southampton Institute, Newman College, Chester College of Higher Education.
Institutions for Medical Training	The Royal Veterinary College, St George's Hospital Medical School.
HE Colleges – Art/Design/ Drama Specialist	The London Institute, Norwich School of Art and Design, Wimbledon School of Art.
HE Colleges – Other Specialist	Greenwich School of Management, Scottish Agricultural College.
FE Colleges	Barnsley College, Bradford College, City of Bristol College.
Private Institutions	SAE Institute Regents Business School, University of Buckingham.

Source: Boliver, 2005.

grant income, amounting to around £1.8 billion in 2006. Not all "research intensive institutions" are part of the Russell Group and in response to this the 1994 Group[3] was set up between those who were excluded. The main difference between the two groups is that the Russell Group institutions tend to have Medical Schools and a scientific focus. The 1994 Group took on a similar remit to the Russell Group of lobbying government, but are less influential because of their smaller sizes and research incomes. The N8 Group is a partnership of eight research intensive institutions including Durham, Lancaster, Liverpool, Leeds, Manchester, Newcastle, Sheffield and York. The initiative was created as part of a joint project between three Regional Development Agencies in the North of England. Unlike the Russell Group that is concerned with lobbying, this group focuses of a series of joint research agendas. A final point on the classification of institutions is that

---

3   www.1994group.ac.uk.

institutions can change their names or indeed composition, and as such, when comparing temporally, these changes must be understood. The definition of an individual institution can change over time, such as the changing of names or the amalgamation or de-merger of institutions. One example was the 2004 merger of the Victorian University of Manchester with the University of Manchester Institute of Science and Technology. Together these institutions became the University of Manchester, and obviously any statistics created to examine pre- and post-2004 will reflect these changes.

Prior Qualification Data

Prior to Higher Education, applicants obtain entry qualifications through the school and Further Education sectors. Compulsory education currently runs up to the age of 16 and culminates in GCSE exams or their equivalent level qualifications, after which the student can study in post-compulsory education or enter full time employment. Post-compulsory education is usually required to gain entry qualifications suitable for Higher Education study. Compulsory education is divided into a series of progressive Key Stages set out in a National Curriculum which represent the standards that students should have met by these points in their education. The Key Stages are as follows:

- Nursery School
  - Key Stage 0 (3–5 years old)
- Primary School
  - Key Stage 1 – Years 1 to 2 (5–7 years old)
  - Key Stage 2 – Years 3 to 6 (7–11 years old)
- Secondary School
  - Key Stage 3 – Years 7 to 9 (11–14 years old)
  - Key Stage 4 – Years 10 to 11 (14–16 years old)
- Sixth Form School/College/Further Education College
  - Key Stage 5 – Years 12 to 13 (16–18 years old)

At the end of Key Stage 2, 3, 4 and 5 students sit a series of tests, which in the case of Key Stage 4 and 5 lead to an individual qualification However in all Key Stages these results are used to construct the annual DCSF schools attainment tables.[4]

There are a series of factors which make data collection, analysis and dissemination for the UK school sector complex. Firstly, since devolution there are different organising bodies for each of the countries which make up the UK. In England this is the Department for Children, Schools and Families (DCSF);[5]

---

4   www.dfes.gov.uk/performancetables.

5   www.dfes.gov.uk.

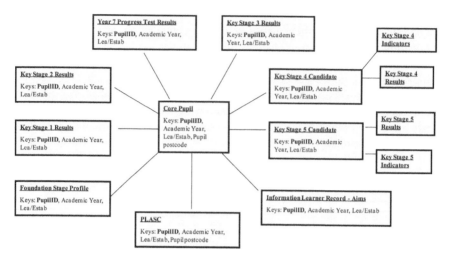

Figure 2.9   The National Pupil Database

Source: Barker, 2006.

in Scotland it is the Scottish Executive (SE);[6] in Wales it is the Department for Education Lifelong Learning and Skills (DELLS);[7] and in Northern Ireland it is Department of Education[8] (DoE). Some curriculum differences occur between the countries. In all countries other than Scotland, Key Stage 4 culminates in GCSE or their equivalent qualification. However, in Scotland, Scottish Standard level qualifications are studied. Similarly, in Scotland at Key Stage 5, Scottish Higher levels are studied; where as in the rest of the UK, A-Levels or their equivalents are taught. A further complication with school data is that independent schools do not have to submit their pupil demographic data to those government organising bodies discussed earlier, thus demographic data for the independent sector is not publically available. In post-compulsory education at Key Stage 5, students progressing from Key Stage 4 do not necessarily study in schools. Some students opt to study in Further Education colleges which have a different organising body, the Learning and Skills Council (LSC).

Those maintained schools and colleges receiving public funding have had a statutory duty to supply data to the DCSF on an annual cycle since 2002 (Jones and Elias, 2006). These data are stored at the DCSF in the National Pupil Database (NPD) across a number of datasets (see Figure 2.9). The linking field which can be used to join demographic data to attainment data at the level of the individual is referred to as the Pupil ID. The records of individuals can also be linked to a

6   www.scotland.gov.uk/Topics/Education.
7   www.new.wales.gov.uk/topics/educationandskills/?lang=en.
8   www.deni.gov.uk.

unique school identifier called a unique pupil number (UPN) to create a series of analysis examining both demographics and attainment within schools. It is these data which are used to calculate school performance tables in England.

The attainment files supplied by the DCSF for research purposes contained attainment results for all students, but personal/demographic data (e.g. unit postcodes) only for those who are included in the Pupil Level Annual Schools Census (PLASC). PLASC is a survey of all students in publicly funded state schools and captures a range of demographic data including:

- Forename and Surname
- Postcode
- Ethnicity
- Free School Meals Eligibility
- Disability Status
- Language of Origin

For access to these data at the unit postcode level, special access requirements were negotiated as these were classified as sensitive data. These data were supplied by the PLASC/NPD User Group at the Centre for Market and Public Organisation (CMPO) in the University of Bristol.[9] As discussed above, demographic data for independent schools is excluded from the NPD, although their attainment data is supplied. For analysis of KS5 there is a further complication, as the NPD contains attainment data for students for who attend the numerous Further Education (FE) colleges managed by the LSC. Additional demographic data for students from these institutions was made available through the LSC. Thus, in order to fully profile the demographic characteristics of the pupils who have attainment recorded in the NPD at Key Stage 5, multiple data sources were required.

Is Higher Education a Public or Private Good?

Earlier it was discussed how participation in Higher Education is growing, however access remains stratified by socio-economic groups. The significance of these access inequalities depends on the extent that Higher Education goods and services are considered as public or private goods by their consumption and benefit limits. Public and Private goods fall within the framework outlined in Table 2.6 (Peston, 1972:13).

Rival goods are those where the consumption by an individual limits the consumption of the good by another individual. The principle of excludability is where only those who purchase a good may enjoy its benefit. Defining a good as purely public or private is difficult, and especially so when the externalities of the product or service are considered. For example, a Global Positioning System (GPS)

---

9   www.bris.ac.uk/Depts/CMPO/PLUG.

Table 2.6   Public and private goods framework

	Excludable	Non Excludable
Rival	Private Good	Common Pool Resource
Non Rival	Toll Good	Public Good

signal could be considered a pure public good; however, the externalities that it benefits only those directly with access to a GPS receiver could limit definition within this category. The Ministry of Defence is probably the only pure public good as it is non excludable, with all people benefiting from the defence, and non rival, as the defence is reasonably equally applicable across all people, perhaps with few exceptions such as extra defence for key figures such as politicians. The extent to which Higher Education in the past, present and future will fit into the Public Good model is a more difficult categorisation. The expansion of Higher Education has allowed more people who may previously have not attended, access to courses of Higher Education, and is pushing the model towards non excludability. However, the extent to which Higher Education is non-rival is debatable, and this depends on whether one is applying for a place at a selecting or recruiting institution. Selecting institutions are those that have capacity short of their demand and as such become selective of those students who are made offers. A recruiting institution is one that can usually accommodate all applicants with the minimal requirements to undertake a chosen course, and as such may, for example, be more pro-active with marketing to fill places. In reality, the extent that institutions fit within this classification will be graduated; however, for the purposes of explaining relative institutional behaviour it provides a useful dichotomy. The equality of access issues discussed earlier undermines the non-excludability principle, as there are certain groups within society who are underrepresented in Higher Education. Therefore Higher Education does not have the characteristics of a standard public good. However, Creedy (1994) discusses that public investment is justified, as Higher Education is at risk of market failure because of the long term horizon of returns to the individual, and that consumption will not be at the socially optimal level without subsidy. The prevailing argument to support the public finance of Higher Education is that it generates externalities that are not directly apportioned to the individual, and as such are beneficial to wider society.

The 2003 White Paper on Higher Education announced that the Higher Education funding gap would be rectified by the introduction of quasi market forces which is contentiously at odds with the government's past image as a public good. Furthermore, the admission by the education minister at the time, Charles Clarke, that not all institutions are the same in terms of the opportunities and life chances that result from courses of study, further adds to this debate (Collins, 2003). This process has been referred as the "marketisation" of Higher Education. However, there is significant evidence to suggest that the public already think of Higher Education in market terms, even with current capped fee prices. League tables are published each year by major newspapers to assess the ranking of particular institutions or courses, and the Research Assessment Exercise grades research output against global benchmarks. Furthermore, the reputation of certain institutions affects the perceived value of courses of Higher Education. For example, a degree from the University of Oxford or Cambridge University will be seen by the majority as more valuable than one from an ex-Polytechnic College.

The true individual financial return from a degree is impossible to empiricise absolutely. However, in 1992, the DfES estimated that graduates would earn over their lifetime £400,000 more than non-graduates (Hodge, 2002a), and the 2003 Higher Education White Paper (DfES, 2003) discusses that this averages around 50% more than non-graduates. These estimated financial benefits have been used in supporting arguments for the introduction of higher variable tuition fees; however they have also been widely criticised. Aston and Bekhradnia (2003) discuss that the 50% earning figure relies on speculative predictions that are impossible to measure accurately. Blasko (2002) adds that the socio-economic background of graduates also influences their relative success in the labour market, and as such the complexity of cultural and social capital would need to be factored into any representative economic degree premium calculation. Therefore, although the quantifiable financial benefit of Higher Education participation is debatable, it seems clear that the skills it provides, and the life chances that result, are all of great benefit to the individual. The economic or societal benefit described in the 2003 White Paper "suggests that there is compelling evidence that education increases productivity, and moreover that Higher Education is the most important phase of education for economic growth in developed countries, with increases in Higher Education found to be positively and significantly related to per capita income growth" (DfES, 2003:58). This is an interesting point, as individual total lifetime productivity is diminished by the three or more years it takes to gain a degree, and as such, this suggests that the extra productivity Higher Education allows to develop is greater than the duration of the course. However, one criticism is that this does not make any reference to the relative productivity benefit for particular courses, assuming the benefit to society is the same across all subject classifications.

In absence of a reduction in real terms, or through re-labelling of institutions categorised as universities, the partial introduction of market forces by the Higher Education White Paper provides an improved but not perfect financial structure which will allow institutions the autonomy to invest in more suitable widening participation strategies. Accepting that institutions have different needs is a step forward, and institution level differentiation within the sector is fundamental to the success of this model. The introduction of market based funding should encourage institutions to embrace their differences through evaluating strengths and weaknesses relative to competitors in order to employ strategies to win market share.

It is argued by some institutions that the current £3,000 fee limit does not go far enough to rectify the deficits in their finances. Sir Howard Davies, director of the London School of Economic stated that the £3,000 maximum fee would only halve the deficit of this "loss making business" (*Economist*, 2004). The majority of selecting institutions charge the full £3,000, and because of their over subscription, the "market" could probably sustain a far higher price without significant loss in applications. Many of the selecting institutions have announced financial participation incentives which have been introduced with these top up fees. Middlesex University has introduced £1,000 bursaries to students who gain places with at least three B grades at A-Level or equivalent (Macleod, 2003). Furthermore, Royal Holloway offered bursaries for a single year's postgraduate tuition after successful completion of their undergraduate studies (BBC, 2004). This supports the marketisation principle in that some institutions are starting to operate more market led recruitment strategies where well qualified students can trade good A-Levels for cheaper admission.

Proponents argue that institutional ability to charge variable fees, combined with decentralised funding sources, will create, in a more traditional sense, a market led Higher Education system that will allow Higher Education institutions to better determine their future (Smithers, 2002). However, there are those who believe that universities are not yet ready to adopt this model, and that our current market-state hybrid system is the worst of both worlds. Scott (2002) identifies several key problems with the introduction of variable fees, discussing how traditional universities, such as those members of the Russell Group, may be inclined to push up fees, not to satisfy market conditions, but to protect their own university brand. These universities may not wish to be seen to charge "bargain" prices as it may reflect on the perceived quality of their products or courses. It is further argued that a positive effect of this could be for universities that currently are at the bottom end of the market, as they may be able to undercut the market leaders possibly through offering reduced fees or attendance incentives, therefore creating new market share. The middle market, made up of the bottom end of the old institutions and the top of the new may be in constant flux. Scott (2002) proposes that some institutions will provide niche courses while others will combine to reinforce their brand. As Phoenix (2003) contends, it is certain "the traditional bilateral relationship between Higher Education and the state is rapidly becoming a multilateral relationship between Higher Education and various external funding bodies". These external funding bodies refer to industry, overseas recruitment and student fees, all of which are adopted in differing mixes to form our current hybrid state-market controlled Higher Education system.

Sir Howard Newby, former Chief Executive of the Higher Education Funding Council for England announced:

> We worry that some institutions might get this wrong [referring to fee price].
> They think they can sustain a £3000 fee across the board, when they will actually
> find they can't. (HEFCE, 2004:1)

This decree came before some institutions announced that they had been considering a strike price to attract targeted levels of students with given levels of attainments and probably also consistent with intended institutional profiles. Sanders (2004) discussed how both Leeds Metropolitan and Bradford Universities were putting together proposals for fees as low as £2,000 across the board.

Lessons from an international perspective teach us that it is very difficult to predict the actual impact that fee levels will have in terms of participation and access. In Australia there has been an increase in participation since fees were introduced in 1989, although opponents do argue that access has contracted during this period. Ramesh (2004) also discusses how graduates from the Indian Institute of Management were dissatisfied with proposals to cut student fees in an attempt to widen access. One student remarked "the government's logic is completely wrong. The fees are high because the facilities here and the Professors are world class and someone has to pay for them".

Market led activities will demand greater intelligence about competitor institutions, their customers' characteristics and, crucially, where they live and are educated. Although a growth area in public policy, there has been relatively little research to date on the specific exploitation of geodemographic techniques, data and tools within the Higher Education context. Two notable exceptions are Tonks (1999) and Tonks and Farr (1995), who examine the applicability of the language and tools of marketing within a Higher Education context. Geodemographic analysis can and will play a key role for institutions to gather essential profiling information, applying tools and techniques more generally utilised by the private sector to target products and services at specific market segments.

Causes of and Solutions to Inequality

Within a context of growing marketisation, access inequalities have remained. Reid (1998) discusses that there are two interpretations of inequality in Higher Education: first, that there is bias in the university selection process; and second, that social class has an inhibitor effect on the perceived availability or benefits of Higher Education. The first of these interpretations was publicly highlighted in 2001 with the case of Laura Spence. Her rejection by the University of Oxford on the basis that she "did not show potential" created a media circus that even involved Gordon Brown, then Chancellor of the Exchequer, who declared it "an absolute scandal". The second of these interpretations relates to how middle class parents "invest all kinds of effort, including significant material resources in developing social capital" (Walker, 2003:172), creating environments where socialisation processes can occur, and creating advantage or disadvantage under certain situations (Bourdieu and Passeron, 1977). Social capital may be defined as the advantage conferred over non-group members through interaction within a network of individuals, who often share similar beliefs or values, and that ultimately lead to greater group-wide economic or social gain. This is not dissimilar to the concept of cultural capital and the two concepts have often been interlinked. Social or cultural capital confers an individual benefit or disadvantage under certain social conditions, such as feeling "comfortable" or enabling interaction with peers within a particular Higher Education institution. Interrelated with ideas of social and cultural capital is the method by which an individual experiences and perceives space. This "space" could refer to an individual perception of whether Higher Education is accessible or restricted. The study of these perceptions is referred to as cognitive mapping, and although this will not be investigated in depth, it is mentioned as an attitude forming framework from which behaviours are measured. Kitchin and Blades (2002:7) discuss that "cognitive maps provide insights into the relationship between people's environmental representation and their behaviour in the environment". Cadwallader (1976) quoted in Kitchen and Blades (2002) discusses how at least three types of spatial decision are influenced by an individual's cognitive map. These are the decisions to stay or go, where to go and finally which route to take. The analogue of this concept to individual decision-making in Higher Education is high, and the extent to which different cognitive maps develop as a result of societal interactions, be this in terms of social class or geodemographic area effects, will be measured in terms of the behaviours recorded in the Higher Education datasets. Knowledge of these behaviours will provide insight into how the cognitive maps of different typologies allow for variable decision-making processes at the area level. Indeed, Krech et al. (1962:20) discuss that the cognitive map is "a partial personal construct in which certain objects, selected out by the individual for a major role, are perceived in an individual manner".

The extent to which one can intervene directly in these access patterns is the realm of social engineering. The introduction of class stratification limits and typology targets are viewed as unethical in the United Kingdom, and any effort to introduce such measures are met with strong resistance by both the institutions and bodies representing the interests of individuals deemed to be disadvantaged by the proposed targets. These issues discussed in general terms by the *Economist* (2004):

> Micromanaging university admissions, as the British government has been trying to do on grounds of class, with targets, quotas, fines and strictures, risks the same consequences as similar American experiments based on racial preference. It humiliates the talented but disadvantaged, whose success is then devalued; it infuriates the talented who are not deemed underprivileged enough and who feel their merits ignored, and it makes universities do a job they are bound to be bad at. (*Economist* 2004:23)

Furthermore, a specific example of resistance to perceived social engineering occurred when UCAS announced that from 2008 the application form would include a question on whether applicant's parents had entered Higher Education. Pat Langham, the president of the Girls School Association, discussed how "favouring candidates whose parents didn't go into Higher Education is artificial and amounts to social engineering" (Lightfoot, 2007). This question was optional, and as such has had limited success.

However, where an institution is unaware of relative performance indicators, and also ways in which various conceptions of capital influence an applicant's prior development and attainment, discrimination can occur as decisions are based on partial information. Perhaps a high achieving applicant to the University of Oxford or Cambridge from a poor performing comprehensive school would not have access to the same literature as an applicant from Eton; therefore the expectation of knowledge of such texts in interview questioning by admissions staff could be perceived as discriminatory. However, this does pose a problem. If a wide literary knowledge is an essential requirement to meet the rigorous demands of a particular course, an admissions tutor without background information may assume a particular candidate is unsuitable, and thus not offer a place. If the admissions tutor had access to contextual information about the applicant's background, this inadvertent discrimination could be avoided, perhaps through supply of introductory readings before commencement of the course.

However, if this information were known, and the desirable quantities of individual groups were specified by policy, then these activities could be considered methods of social engineering, which is attempting to compensate for the failures of state education and social policy by crude adjustments to institutional acceptance profiles. As such, when examining those influences on applications and acceptances within Higher Education, it is important to seek to accommodate such considerations. A key aim should be to effectively extend participation to those segments in society whose participation is currently disadvantaged by internal and external social or cultural values. However, incorporating these ideas into what is historically a tiered applications system will not be without controversy. Those schools that have always sent their pupils to particular universities will resist measures that would result in these patterns changing. Pauline Davis of the Girls Schools Association suggests "it will be difficult, if not impossible, for many of our students to demonstrate exceptional performance in context since the pupils who attend our schools achieve such high standards" (BBC, 2002). This echoes the sentiments of the Headmasters and Headmistresses Conference[10] that represents the views of 250 leading public schools. It produced an investigation in 2002 that showed how, in the worst case, 80% of their pupils were being rejected without interview on certain courses in Russell Group Institutions, claiming that this was a result of positive discrimination policies in these universities (*Guardian*, 2002). Grimson and Dobson (2002) agree, arguing that numerous universities have introduced schemes to increase the total number of state school students without increasing the total number of students, therefore squeezing applications

---

10   The Headmasters' and Headmistresses' Conference (HMC) represents the Heads of some 250 leading independent schools in the United Kingdom and the Republic of Ireland (www.hmc.org.uk).

from independent schools. However, these criticisms ignore evidence to suggest that independent school pupils gain lower degree scores than their state educated equivalents. Allison (2002) discusses how an eight year study of every graduate in the UK revealed independent school students had an 8% lower chance of obtaining a first or upper second class honours degree in comparison with a state school educated pupil sharing the same A-Level results.

Professor Steven Schwartz published the final version of the Admissions to Higher Education Review in September 2004, which has investigated in depth issues of equality in Higher Education admissions. In it he suggests "school type tends to distort the predictive or signalling ability of prior attainment" and that "school performance may also affect the predictive ability of prior attainment" (AHERG, 2004:45). This does raise an interesting question as to whether school type is a direct or indirect indicator of social capital formation. If school type is considered a direct indicator of social capital formation, then attendance will lead to a greater advantage when applying for Higher Education, given that this is a usual and supported course of progression for individuals within these groups. This may occur by being offered better admissions advice when applying, or could be a self-reinforcing phenomenon with each successive generation attending Higher Education, assuming in this social capital model that the perceived socio-economic benefits outweigh the costs. If the school attended is used as an indirect indicator of social capital formation, it may be that the applicant would have made attempts to enter Higher Education independent of whether they attended a particular school type. The school may only reinforce the decision or confer better chances of application success rates. Therefore an underlying focus of the investigation throughout this book is on the nuances of these measures and attempts to generalise their relative importance.

There has been a shift in focus by the media from purely identifying those institutions which are deemed "fair" and "unfair" in their admission practices, to issues caused by the extension of access into underrepresented groups. Hall (2001) discusses that cost of non completion of programmes of Higher Education in 1995–1996 was around £91.5m. In 2002, this figure was re-estimated for the current year at £150m, and the causation attributed to recruiting poorly prepared students (Clare, 2002). The study showed that students from lower socio-economic groups had lower A-Level scores, were applying through clearing and were more likely to drop out of courses. Thomas (2002) discusses how the main factors influencing retention rates include:

- Unprepared for Higher Education
- Prior Academic Experiences
- Institutional Expectations and Commitment
- Academic and Social Match
- Finance
- Family Support
- University Support

The first two points relate to how prior academic experience in the type and nature of qualifications creates different support requirements between groups entering Higher Education. The third influence is the extent to which an individual has their expectations met by an institution. The fourth influence on retention is How both academically and socially an individual perceives themselves as part of an institution. Finance also affects retention, and in particular the extent to which a student feels they are undergoing financial hardship. Family support and university support are related to the extent to which an individual feels they are emotionally and academically supported in their choice of institution and course. Retention studies have investigated this within the context of individuals and the social class to which they belong. As individuals live and interact within areas, there is a plausible assumption to be made that these effects should also be of influence on retention.

Pre-Higher Education Performance

The measurement and interpretation of access rates in Higher Education are often made in ignorance of how attainment stratification is embedded far earlier in the applicants' educational histories. However, in the 2003 Higher Education White Paper (DfES, 2003:68) the DfES has accepted that "the single most important cause of the social class division in Higher Education participation is differential attainment in schools and colleges". Furthermore, Leathwood and Hutchings (2003:137) discuss that prior attainment fits within class profiles and that "working class pupils continue to do less well educationally than their middle class peers" progressively throughout their educational careers. They set out possible reasons for the attainment gap as:

- Poverty
- Family Expectations
- Classed Assumptions about Ability
- Gendered Assumptions about Ability
- Race Assumptions about Ability
- Cultural Capital
- Parental Involvement in Schooling
- Cultures and Practices of the Educational Institutions

However, these reasons fail to mention neighbourhood effects, which could be very influential, as most of the processes above are embedded in different geographies across a broad number of scales.

Selecting institutions are faced with a further achievement problem when selecting the most suitably qualified candidates, in that the number of high achievers is growing. A number of top institutions have introduced extra measures to aid candidate selection such as entry tests (Ward, 2003), and calls for detailed publication of candidate A-Level module results (Hackett, 2003). Nationally, UCAS have replaced the A-Level points system with a tariff score, which incorporates a plethora of other qualifications under the premise that offers will be given in a tariff points, not the previously used A-Level points. To illustrate an extreme example, a candidate can have 360 tariff points by possessing three A grades at A-Level, or, could have an A in the practical and theory CACHE Diploma in Child Care and Education. For a selecting university this would pose a problem, as the "equally qualified" candidates may have the tariff requirements for a 360 point course, but in reality would they be suitably qualified? Ryan (2004:13) discusses that "for almost any course, what a student has done in particular areas is more important that whether they have 120 or 140 points".

The Tomlinson Review published in October 2004 sets out recommendations that were designed to dramatically alter the pre-HE curriculum. This review was justified in Higher Education terms as follows (DfES, 2004):

- Too few young people continue learning beyond compulsory schooling
- It has become increasingly difficult to differentiate between high achievers
- Too few young people have the right skills
  - Communication
  - ICT
  - Number
  - Research Skills
- Too few vocational qualifications meet the needs of learners, Higher Education and employers
- The system is confusing and unclear

The Tomlinson Review proposed that these problems are solved within the new three-stage diploma framework outlined in Table 2.7.

Table 2.7   An outline of the three stage diploma framework

Diploma level National Qualifications	Framework level	Existing national
Advanced	Level 3	Advanced Extension Award; GCE AS and A-Level; level 3 NVQ; equivalent qualifications
Intermediate	Level 2	GCSE at grades A*–C; intermediate GNVQ; level 2 NVQ; equivalent qualifications
Basic	Level 1	GCSE at grades D–G; foundation GNVQ; level 1 NVQ; equivalent qualifications
Entry	Entry	Entry Level Certificates; other work below level 1

Source: DfES, 2004.

However, despite these propositions, the Swartz diploma classification was not implemented and instead the current GCSE and A-Level System were kept. A recent development to tackle the issue of discriminating between the top achieving students is that the Qualifications and Curriculum Authority have announced that A-Levels are to be given an extra grade of A*, where students must attain at least 90%.

Conclusions

This chapter has introduced some of the literature and arguments that underlie and provide context for the empirical analysis presented throughout this book. Higher Education is in a sustained turbulent period of growth in which it has transformed from a minority to a mass market system, with very rapid growth over the last 50 years. Various initiatives have created numerous organising bodies, each with their own data and collection mechanisms. The integration of these data is currently partial at best; there exists overlap within the data collection mechanisms and for some sectors, such as independent schools, demographic data are not collated nationally. Funding for Higher Education is presently improving, although there remains a legacy of underfunding which has left many institutions with large deficits. In a mass model of Higher Education, basic marketisation has been actioned through the introduction of variable fees as a method of rectifying these deficits and sustaining an internationally competitive sector. However, this chapter has begun to highlight ways in which broadening access to Higher Education in terms of absolute numbers may not have the effect of extending access to all groups equally.

Chapter 3
Socio-Spatial Differentiation

The Science of Classification and Taxonomy

Longley et al. (2005:11) discuss how "information systems helps us to manage what we know by making it easy to organise and store, access, retrieve, manipulate and synthesise, and apply knowledge to the solution of problems". Because the world is complex, humans have an intrinsic desire to classify reality and to seek an ordering framework through which information as perceived may be assembled and understood. Bowker, et al. (1999:1) discuss that at a basic evolutionary level "to classify is human" because "human physical abilities are limited, so the amount of information provided to us is constrained by our ability to see" (Weinberger, 2007:4). Humans rarely, if ever, perceive every detail of reality, and as such, create internal cognate models through codifying observations using appropriate levels of detail. These processes are formalised in Psychology through schema theory which describes the construction of those "mental representations which are used during perception and comprehension, and which evolve as a result of these processes" (Anderson, 1977:418). Rosch et al. (1976:382) further posit that these categorisations "are not arbitrary but highly determined", that is there are basic and common categories of understanding between all humans. These occur because:

> The world is structured because real world attributes do not occur independently of each other. Creatures with feathers are more likely also to have wings than creatures with fur, and objects with visual appearance of chairs are more likely to have functional sit-on-ableness than objects with the appearance of cats. That is, combinations of attributes of real objects do not occur uniformly. Some pairs, triples, or ntuples are quite probable, appearing in combination sometimes with one, sometimes another attribute; others are rare; others logically cannot or empirically do not occur. (Rosch et al. 1976:383)

Rosch et al. (1976:383) define a category as "a number of objects which are considered equivalent", and a taxonomy as "a system by which categories are related to one another by class inclusion". As suggested by Schema Theory, taxonomy of categories therefore provides the mental framework through which humans can understand the world. This is exemplified by Davis (1991:21) who relates a schema of an egg to a series of related categories and taxonomy (see Figure 3.1).

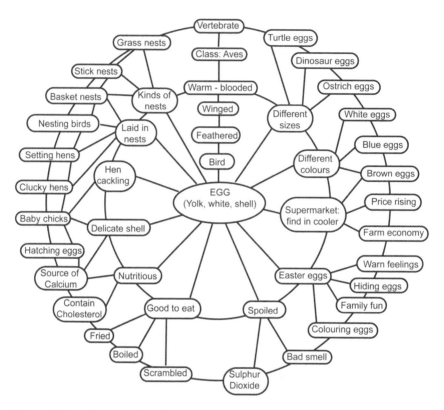

Figure 3.1   An example schemata of an egg

Source: Based on Davis, 1991:21.

These cognate generalisations mirror the competing needs of ideographic and nomothetic science; that is the tension between analyses of the uniqueness of objects versus generalisation of processes. Longley et al. (2005:14) discuss how the technology of GIS, for example, can be used to "combine the general with the specific" through capturing and implementing general knowledge in software while maintaining a database which represents specific knowledge. Within this context, *appropriate* classification can be considered as a problem solving tool which software may use to organise those tacit attributes on the uniqueness of spaces, in order that processes under observation may be understood through empirical observation. The understanding of unique events and processes has been framed as theory formulation by Bracken (1981:112) who suggests "[a] subject without theory is all fact, because facts, or observations, are then only what is *believed* to be correct". This process of theory formulisation is shown in Figure 3.2.

Figure 3.2   The process of theory formulisation

Source: Bracken, 1981:113.

Classification forms an integral component in this process, allowing observations to be summarised and defined before they are conceptualised and later abstracted into theory; however, classification can be refined and improved over time. Bowker, et al. (1999:11) state that "a classification is a spatial, temporal, or spatio-temporal segmentation of the world", and as such the extent to which segmentations can be constructed from fixed natural categories is debatable, as temporal shifts in our understanding of organisation can adapt classification schema to fit a given purpose or agenda (Bowker et al., 1999). The most extreme representation of this definition contends that "everything is miscellaneous" (Weinberger 2007), such that objects can be sorted and classified into multiple different organisations depending on the needs of the user. This concept is explored by Weinberger (2007) in relation to the Internet through his detailed study of modern classification schema, such as the use of "tags" to organise photographs on websites such as Flickr[1] or the development of Folksonomy (Vanderwal, 2007) to categorise blog posts into user defined taxonomies. Despite the seemingly infinite fluidity of informal classification now presented on the Internet, formalised classification has been created throughout history across a range of disciplines as a platform upon which to build scientific knowledge through a shared and common understanding. These will not be reviewed here as adequate coverage already exists elsewhere within the relevant literature (see Weinberger, 2007; Blunt, 2001; Bowker et al., 1999; Lakoff, 1987:92; Borges, 1975:108). The usefulness of classifications in the science of problem solving ideally requires that they are created objectively; however, like those cognate schema discussed earlier, many historically formalised classifications have been recognised to be inherently subjective, often devised by a single or small group of people. However, with many modern classifications, methods have been developed to improve upon their scientific rigour, be this through the use of the kinds of automated clustering algorithms explored later in this book, which objectively seek groups within large multidimensional datasets (see Everitt, 1974), through the introduction of representative scientific review bodies to oversee classification amendments (e.g. International Astronomical Union[2]), or the open dissemination of all data and methods which have been used to

---

1   www.flickr.com.

2   The International Astronomical Union (IAU) was founded in 1919. Its mission is to promote and safeguard the science of astronomy in all its aspects through international cooperation. Its individual members are professional astronomers all

construct a classification (Vickers and Rees, 2007). Within this framework of best practices, classifications are used in this book to provide an organising concept; however, for these classifications to be effective in the context of applications to Higher Education, they must assimilate appropriately defined concepts from existing education theory.

Classification and Educational Concepts

The purpose of formalised classification is to improve our shared understanding of the world by providing a simplified organising framework which helps us understand the complexity of reality. However, "knowledge about how the world works is more valuable than how it looks" (Longley et al., 2005:13), and as such, it is important to embed those behaviours which are classified within an encompassing and appropriate explanatory framework. Within the scope of Higher Education and educational research, both human and social capital have been used to explain and explore why certain recorded behaviours may become manifest.

Higher Education and Capital

In the modern socio-spatial classifications that will be introduced later in this chapter, individuals are aggregated, using their home locations, into a typology based on the average characteristics of the people within areas in which they live. Coleman (1988:S109) discusses how an individual's "background is analytically separable into at least three different components: financial capital, human capital, and social capital". Of these capital constructs, financial capital is perhaps the simplest to address within socio-spatial classifications as it represents the relative advantage gained by possessing increased income or wealth. Using a broad definition of income and wealth financial capital appears to be accounted for very well in commercial socio-spatial classifications, where numerous consumption data are used as input variables. For example, in the literature supporting the Mosaic classification from Experian (Experian, 2007) the following variables are included:

- Credit Behaviour
- Bad Debt
- Shareholdings
- Directorships
- Property Value

---

over the world, at the PhD level and beyond, and active in professional research and education in astronomy (www.iau.org).

However, there are complexities when linking behaviours resulting from differential financial capital if distinction is made between income and wealth. Income broadly relates to the funds gained through receiving a salary. However, after those deductions relating to living expenses, these become disposable income. Wealth is a function of the total value of the assets that an individual owns. For example, an ex-housing association tenant may have bought their home from the local authority in the past, and through inflation they may now own an asset which gives them a high level of wealth: however, their income may have remained moderate. Furthermore, Friedman (1957) hypothesises that individuals may possess a concept he termed as "permanent income". This related to how income and consumption behaviour can be balanced between both permanent and transitory income. Transitory income is the fluctuating income people may receive weekly, monthly or annually. However, permanent income is based on those long term earning projections that an individual may expect to receive. Friedman (1957) argues that consumption behaviour only changes if an individual believes their permanent income will change over the long period, i.e. independent of transitory income. Thus individuals can have varying degrees of wealth, income, and disposable income, and the combination or balance of these concepts will influence how they may behave in a given situation.

Coleman (1988:S100) defines human capital as follows:

> human capital is created by changes in persons that bring about skills and capabilities that make them able to act in new ways.

This concept is more difficult to relate to socio-spatial classification as it does not directly link to tangible assets, e.g. household income. However, human capital can be considered as an enabling framework through which greater financial capital can be accumulated. In the context of attending Higher Education, differences in human capital accumulation might be accommodated within a socio-spatial classification, using indicators of variable educational attainment in post-16 qualification. Reay et al. (2005:21) highlight through an extensive study of access to Higher Education that "an individual's ability to deploy knowledge, skills and competencies is powerfully classed". Therefore, it is likely that between group differences in both physical and human capital might be used to identify groups that are able to leverage different advantages, to different extents, and in different ways.

Social capital is a further type of capital to which Coleman ascribes a more complex definition:

> Social capital is defined by its function. It is not a single entity but a variety of different entities, with two elements in common: they all consist of some aspect of social structures, and they facilitate certain actions of actors-whether persons or corporate actors-within the structure. Like other forms of capital, social capital is productive, making possible the achievement of certain ends that in its absence would not be possible. (Coleman 1988:S109)

Thus social capital "refers to features of social organization such as networks, norms, and social trust that facilitate coordination and cooperation for mutual benefit" (Putnam, 1995:66). Reay et al. (2005:21) describe social capital as "generated through social processes between family and wider society and [...] made up of social networks". Thus, between socio-spatial classification groups which may pertain to different levels of social capital, different behaviours could result through the exploitation of these social networks which enable members to acquire greater social returns. This is exemplified by Edwards et al. (2003:20) who contend that "the literature relating school choice to social networks suggests that the switch from administrative allocation to a 'parental choice' system extends the role of social networks, without necessarily helping to build social capital further. The reproduction of class and community differences is enhanced by the ability of parents to use their power in the education market to shape their children's future milieu".

Cultural capital is a related concept attributable to Bourdieu:

> The notion of cultural capital initially presented itself to me, in the course of research, as a theoretical hypothesis which made it possible to explain the unequal scholastic achievement of children originating in different social classes by relating academic success, i.e., the specific profits which children from the different classes and class fractions can obtain in the academic market, to the distribution of cultural capital between classes and class fractions. (Bourdieu, 1986:243 quoted in Reay et al., 2005)

Thus, in socio-spatial classification, groups with high cultural capital may express certain advantages over those groups with lower cultural capital under specific conditions. For example, entry to university to study a degree in Music may require knowledge of classical pieces with which a student from a socio-spatial group with high cultural capital may have greater familiarity. Swartz (1997:76) describes an ethnographic study conducted by Bourdieu in the French school system, stating that "French school teachers reward good language style, especially in essay and oral examinations, a practice that tended to favour those students with considerable cultural capital who in general are from privileged family origins". In a UK context, Sullivan (2001) demonstrated that cultural capital transmitted within the family home has significant effect of performance in GCSE examination results. However, Sullivan (2001:893) also suggests that once cultural capital had been controlled for in her study, social class largely explained variations in attainment, and as such "cultural reproduction can provide only a partial explanation of social class differences in educational attainment".

Bourdieu and Wacquant (1992) argue that the various forms of capital can be built together, e.g. social and cultural capital could be accrued by purchasing a private education. The process of stratification and exchange can be summarised by a further quote from Bourdieu:

Those with lots of red tokens and few yellow tokens, that is lots of economic capital and little cultural capital will not play in the same way as those who have many yellow tokens and few red ones…the more yellow tokens (cultural capital) they have, the more they will stake on the yellow squares (the educational system). (Bourdieu 1993:34) quoted in Reay et al. (2005:22)

The existence and transformation of the various constructs of capital aim to satisfy the achievement of certain criteria which are "believed to be instrumental in facilitating or blocking the achievement of goals" (Rosenberg, 1956 in Krech, 1962:181). Thus, the aggregate behaviours of people as recorded in empirical analysis of large datasets are related to the differentiation in common attitudes that exist within socio-spatial classification groups. It would be presumptuous to imply that all people within society share common aspirations to improve their relative position in the social hierarchy as generally perceived and measured (e.g. by income, size of house or location), yet there is a body of literature which examines social mobility within this context. For educational markets, this is a highly relevant topic, as it relates directly to our ability to identify those groups in society that may miss out on certain life chances through a restricted ability to compete for the advantages sustained from a quality education.

Social mobility is related to the extent that different groups within society can move upwards within or between the various "networks" and "structures" referred to by Putnam (1995), Coleman (1988) and Reay (2005) in their definitions of social capital. Sometimes referred to as intergenerational mobility, it is defined by the "extent to which a person's circumstances during childhood are reflected in their success in later life" (Blanden et al., 2005:2). An individual's position within such a "hierarchy" of groups is highlighted in some of the earliest definitions of social class, as exemplified in a well known quote by Max Weber:

A "social class" makes up the totality of those class situations within which individual and generational mobility is easy and typical. (Weber 1920:302)

This implies that people exist within a "class" context, and mobility into a different "class" is difficult and atypical. The mapping of these concepts into the context of school and university education has been illustrated by a recent Sutton Trust report.[3] This 2007 study entailed surveying 500 people at the top of their fields in Law, Medicine, Journalism, Politics and Business. Elliot-Major (2007) found that 53% of these leading figures across the five domains had been educated at independent schools, which account for just 7% of school-age children educated in this way nationally. Furthermore, of those who had been to university in the UK, 47% had studied at Oxford or Cambridge (Oxbridge) Universities. Elliot-Major (2007) suggested that "[w]e are still to a large extent a society divided by wealth, with future elites groomed at particular schools and universities, while the educational

---

3   www.suttontrust.com/.

opportunities available to those from non-privileged backgrounds make it much more difficult for them to reach the top." The extent to which industry structures are overrepresented by those who have received the most privileged education provides evidence not only of differentiation by "wealth" but, more broadly, the possible influence of social, human or cultural capital. Thus, it is suggested that possession of increased financial capital broadens school choice, whether through the ability to pay for a private education or freedom to move into areas that are in close proximity to the best schools. Lampl (2007) suggests that "[t]he first priority should be to improve our underperforming state schools, but we also need to recognise that we have a socially selective school system. [...] The top 20% of our secondary schools – independents, grammars and leading comprehensives – are effectively closed to those from non-privileged backgrounds", which presents significant issues of social mobility for those living in less affluent neighbourhoods. A further report commissioned by the Sutton Trust on social mobility concluded "[i]ntergenerational mobility fell markedly over time in Britain, with there being less mobility for a cohort of people born in 1970 compared to a cohort born in 1958 [...] part of the reason for the decline in mobility has been the increasing relationship between family income and educational attainment between these cohorts. This was because additional opportunities to stay in education at both age 16 and age 18 disproportionately benefited those from better-off backgrounds" (Blanden et al., 2005:2). An alternative view to this study has been presented by Smith (2007) who suggests that some of the empirical findings may be misleading, and that the study actually "suggest quite a high degree of social mobility". For the two cohorts of people that were tracked in the study, Smith (2007) suggested that the report actually illustrates that, in the UK, 63% of those born into the lowest quartile of households as measured by income had progressed up the scale, and that in the 1958 cohort this was a similar figure of 69%, marginally better than in 1970. Thus, he suggests that the results actually show a much lower reduction in social mobility than presented by the conclusions of Blanden et al. (2005). Smith (2007) also makes the point that in the 1950s "class" was more closely associated with the grades in an economy led by manufacturing, and that social gradation was then more about manual versus administrative functions. Thus, the Weber (1920:302) notion of class, introduced earlier, where "individual and generational mobility is easy and typical" refers directly to those employment opportunities available at the time. Therefore, it would be highly unlikely for a son of a factory worker to become a lawyer, for example. At this time, assessment of progression between groups would be reasonably obvious; however, in a post-industrial society with a fragmented job market, these differentiations become substantially more difficult to interpret (Longley and Webber, 2003). Thus within this context of social measurement, the following section considers how those concepts relevant to policy issues and outcomes in Higher Education may be measured.

Social Measurement, Classification and Indicators

Social measurement relates to the construction of both classification and indicators which can be used to assess changes in society. Whereas classification implies a multivariate categorisation of objects into discrete boundaries or hierarchy, indicators differ in that they attempt to measure rates in specific conditions either directly or indirectly. Indicators can be either univariate or multivariate in their composition. For example, changes in population characteristics could be measured directly by an individual level national census or indirectly through inferences derived from various data sources such as GP registers, household waste volume or school registrations. However, classification and indicators are not mutually exclusive and, as discussed in the previous section, indicators are sometimes used as input into socio-spatial classification such as those previously mentioned indirect measures of "wealth" that a number of geodemographic companies use as input to their classifications. Additionally, classifications are sometimes used to create local indicators of predicted behaviours (Harris et al., 2005). The complexity of a multidimensional classification, in terms of the mixture and types of indicators that it contains, the diversity of its sources, the different scales of measurement and the varying spatial resolution that it entails, can raise issues that are problematic in public sector applications and increase the need for methodological openness.

Multivariate Indicators in the Public Sector

Considerable research into the classification of populations for public sector applications has been carried out in the past. There has been work in health on the creation of numerous multivariate deprivation indices, with a range of applications in targeting provision and weighting performance. Indicator measures include the Jarman Index (Jarman, 1983), Townsend Scores (Towsend and Beatie, 1988) and the Carstairs Index (Carstairs and Morris, 1991). In each of these classifications, there is a priori reasoning to support the linkage of a concept (specifically, in these instances, the relationship between GP work load and deprivation) to the different sources of data used in construction of the indicator. Each of these classifications is a composite of different Census variables, scaled and grouped in order to measure dimensions of deprivation. The Jarman index, for example, is created from the following variables:

- Unemployment
- Overcrowding
- Lone Pensioners
- Single Parents
- Born in the New Commonwealth
- Low Social Class
- One Year Migrants

The exact calculation using the 1991 Census variables by their standard codes can be found on the Census Dissemination Unit (CDU) website.[4] The original policy objective for the index was to assign extra funding to reflect the workload of general practitioners in deprived areas; however the ability of the index to achieve these aims has been questioned (see Carr-Hill and Sheldon, 1991). Further criticisms of the Index include the spatial variability in performance (Talbot, 1991) and the comparison of the classification to the performance attained using direct socio-economic information (Marsh et al., 2000). Townsend scores were developed for the Northern Regional Health Authority as a further measure of material deprivation, and use the following Census variables:

- Employment
- Overcrowding
- Non Car Ownership
- Non Home Ownership

Although Marsh et al. (2000:630) found "the association between Townsend score and health status was strong enough to be of practical importance", the variables included in creating the summary score should only reflect aspects of deprivation which are uniform across all areas. Car ownership in a dense urban area such as London, typically would not indicate deprivation because other modes of transport are used more prevalently. The Carstairs index was developed subsequently from the Townsend scores by the analysis of Scottish health data. This new classification was sought to reflect certain elements of deprivation which were believed to be unique to Scotland. The following variables are included in the Index:

- Unemployment
- Overcrowding
- Non Car Ownership
- Low Social Class

Since the year 2000, two re-workings of a related, more generic, deprivation indicator have been created for England. The most current of these is the 2007 Index of Multiple Deprivation (IMD).[5] The IMD is disseminated at Lower Layer Output Area and is designed for a range of public sector applications. The classification is created from seven different domains including:

---

4   www.census.ac.uk/CDU.

5   The IMD is supplied by a department of the government called Communities and Local Government (Formally the Office of the Deputy Prime Minister). The classification is available from: www.communities.gov.uk/communities/neighbourhoodrenewal/deprivation /deprivation07.

- Health Deprivations and Disability
- Employment
- Income
- Education, Skills and Training
- Living Environment
- Barriers to Housing and Services
- Crime

In addition to scores and rankings for these seven domains, an overall score is created as a measure of general deprivation. Although freely available to use, there is little evidence that either the aggregate classification, or the education domain have been used in peer reviewed education literature to the extent that is has been adopted in health. HEFCE has made some effort to introduce spatial classifications into its funding models with the introduction of the Participation of Local Areas (POLAR) classification in 2003.[6] This classification was derived from a demographic study of young (18–19 year old) participants in Higher Education between 1994 and 2000. Rates were derived at the Ward level and these scores disseminated as a series of quintile bands.

Some Examples of Measurement in Education

In educational applications there are various methods which are or have been used past and present to transform data into information (Longley et al., 2005) including classification, direct and indirect indicators. Investigations into Higher Education participation have most prevalently used individual level occupational based classification. Archer et al. (2003) review these classifications to include the National Statistics Socio Economic Classification[7] (see also Rose and Pevalin, 2003), the Socio-Economic Group and the Registrar General's Classification. These formal classifications group individuals into a classification category according to occupation. For applications in Higher Education the use of occupational classifications should cause concern. When assigning these variables before the age of 21, UCAS assigns occupational categories based on applicants' parental occupation, whereas after the age of 21, these are assigned on the basis of applicant occupation. Thus, within this classification parental occupation is used as an indirect indicator of the student's background, whereas a more appropriate measure could perhaps be derived using a neighbourhood level classification that directly reflects the milieu in which the individual grew up. Since 1994, the UK Government has published school performance tables for secondary school GCSE and A-Level results, which additionally were supplemented with primary school attainment data in 1996. These school performance data directly measure

---

6   The full methodology for constructing POLAR can be found at: www.hefce.ac.uk/pubs/hefce/2005/05_03.

7   www.statistics.gov.uk/methods_quality/ns_sec.

attainment in exams taken by students within those schools and allow comparison to be made against equivalent performance both within the local authority and also the national average. National coverage school data do not include parental occupation and as such, in educational analysis, socio-economic differentiation is often implied through analysis of free school meal rates within schools as a substitute for a level of deprivation (see Shuttleworth, 1995). In Higher Education, performance indicators are constructed for widening participation, retention rates, research output and employability of graduates. These are created from a range of both direct and indirect indicators taken from those attributes collected on students studying within Higher Education in the previous academic period.

Static Classification and Dynamic Processes

Social measurement has been discussed thus far in terms of static states of behaviour. However, these measured events do not occur in temporal isolation, and as such social measurement can be considered by the extent that classification or indicators capture these dynamic processes.

This is a relevant concept to the urban land use models developed in early 20th century Chicago. Robert Park and colleagues applied ecological concepts to explain observed segregation and patterns of social interaction within Chicago. Knox (1987) discusses how these "natural" processes included dominance, segregation, impersonal competition and succession. Park proposed that the fundamental principle of competition between individuals occurred through the operation of markets and resulted in patterns of land use and rent, and therefore, the segregation of people into distinctive areas. Burgess (1925) used these ideas to model the processes of neighbourhood differentiation, developing his famous concentric ring model (see Figure 3.3). In this model, the city grows outwards and differentiation between changing zones is established by the ecological processes of invasion and succession, or in other words, expansion and land use conversion. The contribution of these studies to the history of urban measurement lies in the paradigm shift that it stimulated away from focused research on specific areas to the establishment of more general statistical relationships between urban systems and their structure (Harris et al., 2005).

Further models that developed out of this same school of thought included Hoyt (1939) who proposed a sector model that was based on land use being divided by a series of wedges that radiate from the CBD to account for the effects of communication lines such as roads (see Figure 3.4).

However, the Human Ecology school subsequently became discredited as the models were applied to other locations outside the case study cities, and Knox (1987:61) states that was later "abandoned in favour of a much modified ecological approach based on the idea of identifying key variables and examining their relationships within an ecosystem or ecological complex". These social area analyses were developed by Shevky and Williams (1949) and involved creating a series of scores for a number of variables including social rank, segregation and

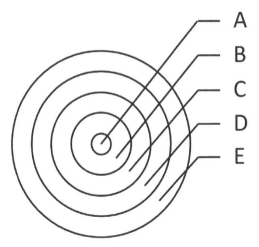

Figure 3.3   The Burgess Land Use model

Key: A= Central business districts, B = Zone of transition, C = Zone of factories and working men's homes, D = Residential homes, E= Outer commuter zone.

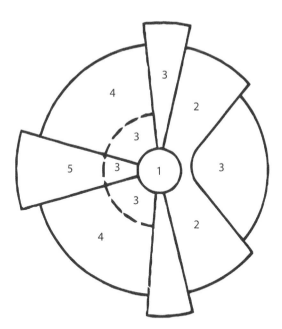

Figure 3.4   The Hoyt Sector Model

Key: 1=Central business area, 2=Wholesaling and light manufacturing, 3=Low income housing, 4=Middle income housing, 5=High income housing.

Source: Hoyt, 1939 in Harvey, 2000.

the degree of urbanisation, all within the urban area of in Los Angeles (LA). Using these three indices, LA census tracts were divided into an 18 Group classification based on how similar the areas were between the three indices. Bassett and Short (1980) discuss how initial reviews of this work pointed out that there was little theoretical justification for the variables that were selected and the indices that were created using them. What became known as social area analysis was later re-presented by Shevky and Bell (1955) as a deductive model: however Bassett and Short (1980:17) note that "social area analysis as theory is little more than a retrospective rationalisation for previous empirical work". Thus these early models, although a progressive step forward in urban theory, were essentially created through a process of inductive theory building, which is "theorising from a mass of observations" (Wilson, 1971:32) and therefore illustrate a weak linkage between concepts and measurement. Thus, it is important for modern measurement techniques to be grounded in theory which can be validated through repeated observation.

Geodemographics, Socio-Spatial Differentiation and Change

Geodemographics represents a multidimensional technique to measure socio-spatial differentiation and change extending as a progression from those urban models in the "ecological tradition" (Bassett and Short, 1980:9). These techniques are essentially deductive models of urban processes which build from those models discussed in the previous section and additionally relate to work Booth (1889) undertook to study poverty in London. Booth classified streets in London by the prevailing socio-economic characteristics of the areas, assigning them into one of seven groups which were colour coded and linked to a series of descriptions which represented a range of social conditions, specifically poverty.

Harris et al. (2005) discuss how the clustering of small areas into groups, using numerous census variables, originated in the late 1970s as a deprivation classification of wards, parishes and enumeration districts in Liverpool. The Office of Population Censuses and Surveys (OPCS) commissioned this research from the Centre for Environmental Studies in London, based on previous small area deprivation studies in Liverpool (Sleight, 1993). Richard Webber is credited with much of this early research (see Webber, 1977; Webber, 1978; Webber and Craig, 1978) and was eventually commercialised by the company CACI who introduced these methods into the private sector under the brand ACORN (A Classification of Residential Neighbourhoods). These data on socio-economic characteristics were thus used to imply consumption characteristics relevant to the private sector. The classification was later re-created at the level of the unit postcode.

These classifications potentially commit ecological fallacy (Robinson, 1950) by design, since they seek to predict individual consumer behaviour using indicators pertaining to areal aggregations. However, the strength of these problems depends on the exact area of aggregation being studied (Martin, 1991).

Furthermore, geodemographic analysis may suffer problems with the Modifiable Areal Unit Problem (MAUP) (Openshaw, 1984) which occurs when the mean attribute values of geographical areas change depending on the number of areas into which a population is divided (Tranmer and Steel, 1998) or when zonal boundaries are moved or modified. Separately or together these scale and zoning effects form the MAUP. This is of potential concern to the construction of geodemographic classifications, since some are built as aggregations from a range of geographical units and none is founded exclusively upon observations pertaining to unique human individuals (Fotheringham and Wong, 1991). This is most recently demonstrated in a geodemographic context by Experian releasing the Mosaic classification for academic use into the National Data Archive. The data supplied were aggregated up from unit postcode to district level. However, Richard Webber notes in De Smith et al. (2007:96, cited as personal communication) as response to criticisms of Geodemographics in relation to MAUP that:

> I have never to come across any real world example of a conclusion being invalidly reached as a result of this hypothetical possibility. (Webber in De Smith et al. 2007:96)

There has been little attempt within the literature to identify the relationship between those concepts of capital formation introduced earlier in this chapter and geodemographic analysis, perhaps with the exception of Webber (2007). Unlike "class" which is a concept traditionally used and understood in the broad sociological literature, geodemographic typologies have only recently begun to seek to provide, or perhaps post-rationalise, a conceptual framework to understand reality (see Parker et al. 2007). The experience of Shevky and Bell (1955) suggests that we have been here before. Drawing parallels between concepts of social stratification derived from sociological literature and labels used in the Mosaic typology, Webber (2007:185) suggests that these "incorporate language which corresponds closely to that used in the discourse on both globalisation and gentrification, and studentification", although he cites only a single source (Atkinson and Bridge, 2005) in making these assertions. The shorthand labels and pen portraits (Birkin, 1995) assigned to the elements of different classes within any typology are undoubtedly useful, yet vivid description does not of itself provide academic rigour, particularly when the methodology for the construction of the commercial classification is closed to external scrutiny. There are indeed worrying similarities between the criticisms that Skevky and Bell (1955) sustained in relation to their post facto rationale of social area analysis and the arguments presented by Webber (2007) in support of geodemographics. Geodemographic analyses clearly have an inductive legacy, and to retrospectively argue that current typologies are founded upon ad hoc strands of social theory is academically misleading. A different route is for classification vendors to re-create future classifications from a more theoretically informed and deductive standpoint.

Within the private sector, geodemographic analysis has conventionally been used to examine both incremental purchases (such as those measured by newspaper readership) and discrete consumption of private goods (such as the propensity to consume holidays of different types). In principle there seems no reason not to anticipate the suitability of these techniques to investigate aspects of consumption of public sector goods and services. However, government statistics have traditionally used data derived from the census, administrative records and sample surveys to allow them to meet the majority of their needs for services targeting in local areas (Longley and Webber, 2003), and for this reason geodemographic analysis has to date been less prevalent within public sector applications. There has been a growing number of applications in health (see Aveyard et al., 2000; Tickle et al., 2000; Stafford and Marmot, 2003), crime (Massimo et al., 2001; Bowers and Hirschfield, 1999, Ashby and Longley, 2005) and education (Tonks, 1999; Tonks and Farr, 1995) that explore the use of GIS and geodemographic analysis to assist in policy and decision making at both the local and national levels. These studies have developed in parallel with government reforms of public services, and indeed, the initiative "Big Conversation" encouraged discussion on replacing a one size fits all welfare system with one of "individual aspiration backed up by strong communities" (Blair, 2003). This shift in the focus of public policy towards individuals and the neighbourhoods in which they live is driving micro level data requirements, and to some extent mirrors the developments seen in the private sector.

A further strand to these shifting interests stems directly from government policies that have encouraged public sector organisations to behave in ways akin to those of the private sector. These convergences can be illustrated by the replacement of government Compulsory Competitive Tendering Policy with the concept of "best value" for the repeat purchase of public sector goods and services. This policy is consistent with public-private collaboration, and as the geodemographic industry has been shown to be successful and highly profitable in the private sector (Harris, 1998), its adoption in the public sector seems to be a logical and progressive step forward. This becomes of growing importance as the geodemographics industry becomes increasingly open to competition in the UK, not least as 2001 Census data are now freely available to download, allowing end users the ability to create their own systems or "value added" by incorporating their own sector specific data. Prior to the 2001 Census, this was also possible, but only by being tied into expensive licence agreements with the small number of census distributors who in turn licensed public sector data from the Office of National Statistics (ONS)/Office of Population Censuses and Surveys (OPCS). The free availability of the 2001 Census data has led to the creation of the ONS Output Area Classification (OAC: Vickers and Rees, 2007) which has an entirely open and public domain geodemographic classification. This classification is seen as a positive step forward; although without the underlying data being refreshed (as in commercial classification) there is danger that this classification could become out of date.

Privacy and Ethics

When attempting any form of categorical classification of areas or individuals, there is the underlying assumption that reality can be accurately represented by the typologies, and debates and dialogues justifiably extend to how the groups were conceived, measured and otherwise created. For classifications used in the public sector this is of particular significance, especially if the application of techniques may affect the real life chances of those who are classified correctly or otherwise. Questions concerning the integrity underpinning the use of geodemographics in the analysis of social processes and stratification often cite Harvey's (1973) concern that these techniques develop knowledge that appears to be true, but that in actuality hides the truth of reality. Furthermore, Sui (1998:662) suggests that these "instrumental approaches generally take an atomistic ontological position in which the social position of the researcher is independent of the knowledge that he or she produces"; suggesting that data led empirical investigations may not be sufficient to represent the complexity or dynamics of real world social processes.

Geodemographic information systems have also been criticised as they threaten privacy in two key ways. First, Goss (1995) describes how mis-specification problems within a database can inadvertently discriminate, even if the use of the data is legitimate in de jure terms. In the context of Higher Education, these effects could be particularly acute, as the consequences of mis-specifying disadvantage in terms of educational services has potentially very serious consequences, and particularly so if life chances are being apportioned. It is therefore essential that, if these indicators are adopted, analysis be conducted into whether they will ameliorate or compound these issues. Goss's (1995) second privacy concern is that data on an individual gathered for one purpose may be transferred to another context without that individual's permission. "Off the Shelf" geodemographic indicators are constructed with legally available data without infringement of this aspect of the Data Protection Act, thus negating this second concern, albeit only in a strict legal sense. However, when these indicators are appended to other data such as university application successes, extra caution must be taken to ensure that the spirit of these safeguards is maintained. Taken together, it is clear that when using either geodemographic or socio-economic indicators, it is important to ensure that the context of the investigation is understood, in order that erroneous interpretations may be avoided. For example, in an investigation into Higher Education participation using the Mosaic geodemographic classification, it may be inferred that an area classified as "Welfare Borderline" has lower Higher Education participation rates because of the restriction created by prevailing lower incomes (assuming that this was identified as a key variable for lower participation). Yet in reality there may be other important contextual factors in these areas such as the types of employment, and it is these that may affect the weighting or importance that is placed upon Higher Education and attitudes to social capital formation. Resulting human capital formation from participation in Higher Education is unlikely to be a simple function of income alone and it is more likely to accrue through a combination of socio-economic processes.

Measuring and Modelling Educational Choices and Decisions

The usability of a classification schema in measuring or modelling educational choices is directly related the homogeneity of the composite groups of applicant/ institution choices and decisions. Individual participation choices fall within the broad categories of whether, where and which Higher Education courses attract the participation of an individual. These choices then intersect with the decisions that an institution makes, such as offer or rejection. In this context education can be considered a commodity, and applicants as consumers. Hensher and Johnson (1981:12) suggest that the "selection decision between commodities which are perceived to be available" is influenced both by physical and aphysical factors. Physical effects on demand could be course grade requirements, and aphysical factors could be that a candidate may perceive a particular institution as "too posh", or as "ivory towered" (Singleton, 2003). Choices may be deemed to entail "selection decisions by an individual between commodities which are perceived to be discrete and which are contained in a relative choice set" (Hensher and Johnson, 1981:11). Commodities are defined at either the intensive or extensive margins. Higher Education provision falls at the extensive margin, as an increase in cost cannot usually result in a partial reduction in consumption in the same way as with a continuously defined product such as the purchase of meat from a butcher. A consumer, in this case applicant, will choose (or not) a course of Higher Education that maximises their utility from a choice set by consuming different levels of attributes from choice alternatives, where utility is the "index of the relative levels of satisfaction associated with the consumption of particular commodities" (Hensher and Johnson, 1981:11). In Higher Education these attributes could include qualification aim, price of the course, or reputation of the institution. An applicant will make joint course and institution choices according to a set of value judgements that maximise perceived personal benefit over cost. Factors that may influence these choices might include:

- Residential Location
- Subjects Offered and Course Structure
- Facilities
- Grades
- Reputation
- Parents and Family
- Friends
- Schools
- Price
- Bursaries

Therefore, using the maximum utility principle an individual choice set may be limited or enhanced by one or combination of these factors. For example, a single mother living in Bolton and wishing to extend her education, may have a choice set constrained to a single institution. However, a final year A-Level student, living in Buckingham and attending a top performing school may have a much less constrained set of choices.

Institutions will also make value judgements as to which students they wish to admit in order to meet their recruitment or widening participation targets. These might include the following variables:

- Personality
- Interview Performance
- Grades and Subjects
- Entrance Tests
- School Type
- School Achievement

The interaction between student and institution choices creates a distribution of acceptance to courses of Higher Education. The effectiveness of a geodemographic classification for use in Higher Education might therefore be a function of its tacit ability to measure the variables which influence these behaviours. As part of a major 39 month research project into educational choices which consisted of around 137 in depth interviews, Ball et al. (1996:104) discuss how socio-spatial differentiation in education arises and is reinforced as a function of the choice process available to different segments of society:

> ...choice is very directly and powerfully related to social-class differences... choice emerges as a major new factor in maintaining and indeed reinforcing social-class divisions and inequalities. (Ball et al. 1996:104)

Ball et al's (1996) use of "maintaining" and "reinforcing" links with the role of education in restricting or enhancing social mobility as shown in literature reviewed earlier. The apportionment of different constructs of capital between people in society creates variability in the choice sets available to them. Gewirtz et al. (1993) introduced an organising framework for these choices of *inclination* and *capacity*. Tooley (1997:218) summarises these two factors as follows:

> Capacity itself has material and cultural dimensions. Material capacity includes the resources to pay for transport to and from school (including private cars and taxis), improved housing, extra-school coaching, private school fees and child-care opportunities. Cultural capital capacity includes having knowledge about and familiarity with the education system, self-confidence, and "stamina-to research, visit schools, make multiple applications and appeal". (Tooley 1997:218)

> Inclination involves the extent to which they were inclined to be engaged with the choice system; higher inclination implies possessing certain "beliefs" about

schools, for example, that they differ in terms of: atmosphere, the "standard of education" they offer, their exam results, the life chances they facilitate, the values they impart, the extent of extracurricular activities, the kind of children that go to them, levels of resourcing, levels of parental support, and the commitment of teachers. (Tooley 1997:218)

From the literature reviewed previously relating to capital, Tooley's (1997) capacity directly relates to the various forms of capital discussed earlier. However, inclination is a more complex attitudinal construct and relates to the engagement and motivation of the different groups which make up society to make educational choices. Furthermore, this conception of educational choices relates to utility maximisation (Hensher and Johnson, 1981), also discussion earlier. Those groups of people as defined by a geodemographic typology are going to have variable capacity and inclination, thus placing them within the Gewirtz et al. (1993) choice matrix as presented by Tooley (1997:219) at different locations (see Figure 3.5).

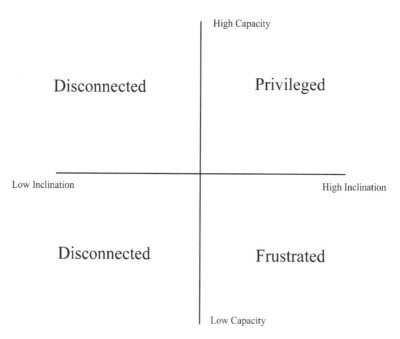

Figure 3.5   Matrix of choice characteristics

Conclusion

The emergent contention of this wide-ranging review is that for a classification to effectively differentiate in educational applications, it must seek to measure the behavioural differences arising from choice restrictions exhibited through variable inclination or capacity. In order to ensure that the concepts discussed throughout this chapter are effectively mapped into homogeneous classification categories, there is perhaps a requirement for bespoke educational classifications or indicators which integrate these behaviours directly through use of the underutilised data that were discussed in the previous chapter. From this discussion, the interim conclusion is that it appears that general purpose classification systems (such as those marketed by commercial providers) can claim no particular status in accounting for the consumption of the various services provided by the *public* sector – with perhaps the single partial exception of Health ACORN.[8] This does not directly include health sector data, but includes other data relating to health outcomes, such as diet information. However, the implied assumption that that the nature of individual use of public services such as education should directly correspond with the ways in which consumers uses private goods and services is problematic, if not fatally flawed.

---

8   www.caci.co.uk/acorn/healthacorn.asp.

Chapter 4
The Socio-Spatial Context
to Higher Education Access

How Can We Understand Access?

This chapter will introduce a series of univariate indicators which enable better understanding of what we mean by access to Higher Education; how distributions may manifest and be appropriately measured. This analysis aims to unravel some of the social and spatial complexities which surround access issues in Higher Education by using geodemographic classification as a multidimensional organising framework through which aggregate behavioural choices can be measured. A series of analyses are presented which relate to a number of important indicators that are highlighted by the literature to influence differential social and spatial neighbourhood access rates including distance (Harris et al., 2007), prior attainment (Leathwood and Hutchings, 2003), course choice (Reay et al., 2005) and age (Archer, 2003).

Consideration of these issues begins to develop a notion of which indicators are to be most usefully included in those tools required for decision support. Linking access indicators to a common framework of geodemographic classification provides stakeholders with a simple method through which those complex behaviours can be measured. Different stakeholders, however, will be interested in different indicators. For example, selecting institutions may be interested in those students from neighbourhoods who supply the best prior qualified, candidates whereas recruiting institutions may have more concern for the distances that students from particular neighbourhood groups typically travel to accept an offer at an institution.

This chapter explores the spatial and social complexities that are present in geographies of access to Higher Education as "it is generally helpful to look at a data set before any models are fitted or hypotheses formally tested"! (Fotheringham et al., 2000:65).

Higher Education Choices and Distance

In the UK, Gridlink®[1] is a joint venture between a series of national organising bodies, including the Ordnance Survey and Office for National Statistics (ONS)

---

1   www.statistics.gov.uk/geography/gridlink.asp.

which enables unit postcodes to be spatially referenced. When UK applicants to Higher Education provide UCAS with their completed application form, the postcode from the home correspondence address can be used to geocode the residences of prospective students. Previously known as the All Fields Postcode Directory (AFPD), and since May 2006 as the National Statistics Postcode Directory (NSPD), these files contain a list of all current and expired postcodes in the UK, along with a series of lookups to corresponding geographic boundaries. The process of appending a spatial reference to a unit postcode involves calculating a population weighted centroid of all the delivery points within each unit postcode. The AFPD and NSPD file available to the academic community through Edina's UKBorders[2] service have a centroid Easting and Northing spatial resolution at up to 1m. UCAS define the location of a Higher Education institution based on the unit postcode associated with the main institutional campus, and as such this can introduce some error into the data from those institutions that have multiple sites. Using the AFPD file, 2004 UCAS degree applicants who accepted a place were geocoded using their domicile postcode. The spatial location of accepting institution was defined in the same way as the home location and the Euclidean distance between these two points was calculated for each applicant. These straight line distances between student domiciles and institutional locations i is calculated using Equation (4.1).

Equation (4.1) – $\quad d = \sqrt{(x_s - x_i)^2 + (y_s - y_i)^2}$

Distance scores can be aggregated and partitioned using a series of other variables available in the UCAS data. This section will examine some of these spatial complexities, demonstrating how courses, institutions and socio-economic factors all influence the distance that an applicant is willing or able to travel to accept a place at an institution. Figure 4.1 shows the median distance travelled by each placed student according to 2001 Census Super Output Area geography. The darker colours indicated longer distances travelled and the dots are the location of UCAS institutions. Where clusters of institutions occur there is a tendency for shorter distances to be travelled, most significantly represented in London (see map inset). From rural areas, as one might expect, students generally travel longer distances.

Although of interest to national Higher Education policy, the aggregate spatial distribution shown in Figure 4.1 does not highlight heterogeneity in distance travelled to different Higher Education institutions, to participate in different courses, or, from different neighbourhood Types. Calculation of distance travelled according to Mosaic Types for all 2004 UK degree acceptances (see Figure 4.2) makes clear the high variability between Types and Groups. These patterns are partially explicable by the spatial distribution of the neighbourhood Types themselves. Accepted applicants from the Group "Rural Isolation" have a higher

---

2   www.edina.ac.uk/ukborders.

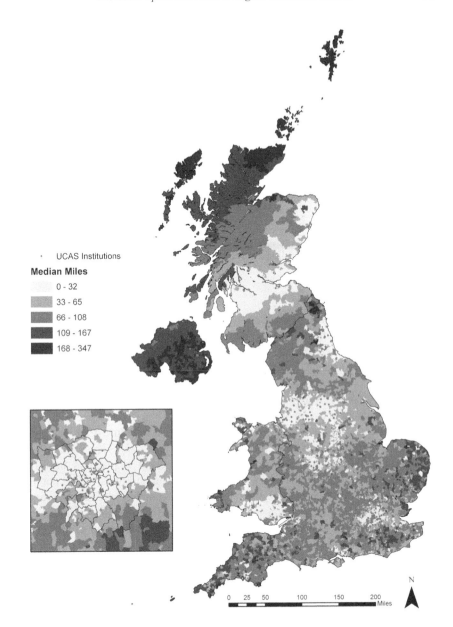

**Figure 4.1   Median distances travelled by students resident in each UK
                Super Output Area**

Source: UCAS Data, 2004.

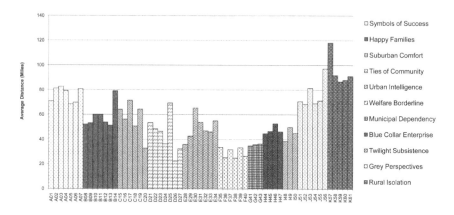

Figure 4.2   Distance travelled to accept a degree place by Mosaic Type
Source: UCAS Data, 2004.

propensity to travel greater distances, as these neighbourhood Types occur in areas distant from the major urban conurbations where majority of Higher Education institutions are located. Another Mosaic Group with a high propensity to travel is "Symbols of Success", which contains the most affluent neighbourhoods in the UK and that are often found in large and prospering cities. These applicants will probably have fewer financial restrictions on their abilities to travel larger distances to accept places at an institution, and with a history of Higher Education more prevalent within the families living in these areas, a culture of moving away from home to university is more likely.

Although the average distance travelled to accept a place is useful in describing the aggregate behaviours of particular neighbourhoods, it does not help institutions to define their relative market areas. For example, Lancaster University has a median acceptance distance of 58 miles, with variability according to neighbourhood Type shown in Figure 4.3. This is a very different profile to the University of Westminster, with a median acceptance distance of eight miles and the neighbourhood Type variability shown in Figure 4.4.

The variability in the overall median distance that is travelled to accept a place is influenced by the proportion of students from different neighbourhood Types that accept places. Therefore, in the example above, the University of Westminster will accept a majority of its students from neighbourhood Types with low propensities to travel.

Further insight can be gathered if the incoming attainment level required for each institution is compared against these distance profiles. Attainment may be measured by an average of accepted applicant UCAS tariff scores. These are shown against median distance scores in the scatter plot in Figure 4.5, where the point labels correspond to UCAS institutional codes. There is a positive correlation indicating that institutions with high average entry requirements also attract applicants who appear to feel able to travel greater distances to accept places. This is confirmed when the average tariff scores are classified by Mosaic Types in Figure 4.6. When this chart is compared with the average distances that applicants travel to accept places (Figure 4.2), the patterns are similar, with those neighbourhoods supplying students who travel further also entering with higher attainment.

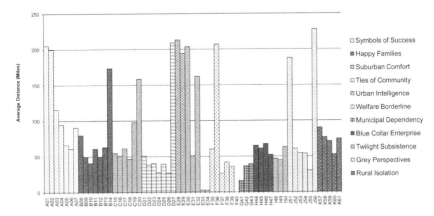

Figure 4.3   Distance travelled to accept a degree place by Mosaic Types at the University of Lancaster

Source: UCAS Data, 2004.

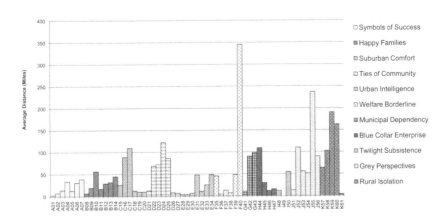

Figure 4.4   Distance travelled to accept a degree place by Mosaic Types at the University of Westminster

Source: UCAS Data, 2004.

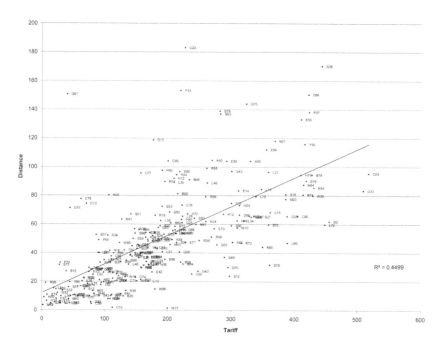

Figure 4.5   A cross tabulation of institutional average tariff scores against average distance travelled to accept degree offers

Source: UCAS Data, 2004.

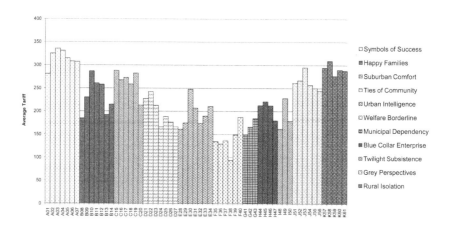

Figure 4.6   Average tariff scores by Mosaic Type for accepted degree applicants

Source: UCAS Data, 2004.

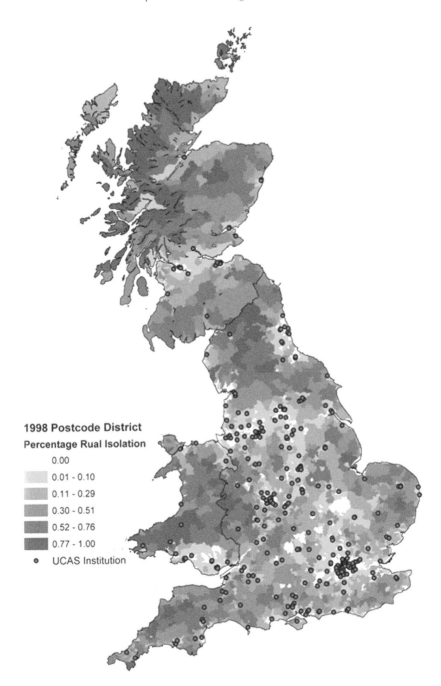

Figure 4.7   Percentage of Mosaic "Rural Isolation" postcodes

Source: Experian Mosaic and Author calculations.

Figure 4.8   Institutions with degree acceptances on courses in Maritime Technology – J6 and Naval Architecture – H5

Source: UCAS Data, 2004.

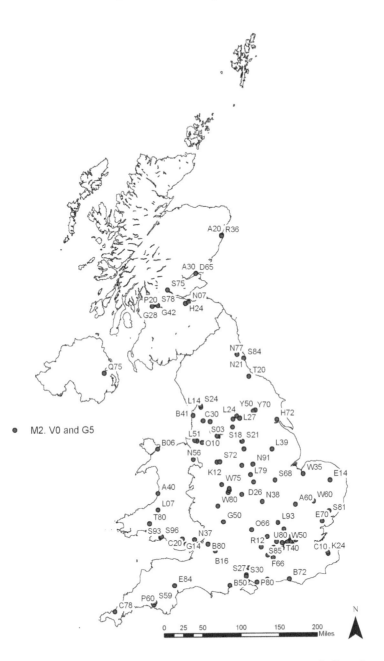

Figure 4.9 **Institutions with degree acceptances on courses in Law by Topic – M2, Historical and Philosophical Studies – V0, and Information Systems – G5**

Source: UCAS Data, 2004.

The variability in distance travelled to accept places by neighbourhood Types and between institutions is influenced by the geographic location of each institution in relation to particular neighbourhood Types. For example, the University of Westminster is located in London where there is a significant proportion of the national tally of "Symbols of Success" neighbourhoods. Although this Group has a tendency nationally (Figure 4.2) to travel longer distances to accept a place, they do not exhibit this pattern at this particular institution (Figure 4.4). To illustrate how different geodemographic neighbourhood Types are spatially represented, Figure 4.7 shows the percentage of neighbourhoods (unit postcodes) coded as "Rural Isolation" within each postcode district.

The dots on the map are UCAS institutions and illustrate how the incidence of different neighbourhood Types varies relative to the locations of different institutions. Furthermore, because the course offerings of different Higher Education institutions are unevenly distributed across space, Higher Education courses also have variable distance profiles. Table 4.1 and Table 4.2 show the top and bottom 20 degree courses

Table 4.1   Top 20 average distance travelled to accept degree courses

Rank	JACS Subject Line	Miles	Acceptances
1	J6 – Maritime Technology	127	153
2	H5 – Naval Architecture	115	113
3	T2 – Japanese studies	113	144
4	T1 – Chinese studies	112	84
5	L6 – Anthropology	110	534
6	T9 – Others in non-European Langs and related	109	897
7	D5 – Forestry	106	61
8	QQ – Combinations within Linguistics, Classics and related	105	498
9	D1 – Pre-clinical Veterinary Medicine	104	873
10	V3 – History by Topic	104	1,113
11	Q8 – Classical studies	103	800
12	VV – Combinations within History and Philosophical studies	98	1,188
13	F7 – Ocean Sciences	98	214
14	C3 – Zoology	97	1,105
15	C2 – Botany	96	23
16	V5 – Philosophy	96	1,327
17	V6 – Theology and Religious studies	94	1,156
18	RR – Combinations within European Languages, Lit and related	92	1,838
19	LL – Combinations within Social Studies	92	2,436
20	F6 – Geology	91	1,344

Source: UCAS Data, 2004.

Table 4.2   Bottom 20 average distance travelled to accept degree courses

Rank	JACS Subject Line	Miles	Acceptances
1	M2 – Law by Topic	32	1,263
2	V0 – Hist and Philosophical studies: any area	32	284
3	G5 – Information Systems	32	3,753
4	BB – Combinations within Subjects allied to Medicine	36	223
5	L5 – Social Work	37	5,330
6	XX – Combinations within Education	37	227
7	X3 – Academic studies in Education	37	3,784
8	X1 – Training Teachers	38	7,636
9	KK – Combinations within Architecture, Build and Plan	38	163
10	F0 – Physical Sciences: any area of study	39	215
11	W9 – Others in Creative Arts and Design	39	539
12	MM – Combinations within Law	39	259
13	X9 – Others in Education	40	163
14	B7 – Nursing	40	5,684
15	P4 – Publishing	42	176
16	B8 – Medical Technology	42	1,669
17	L4 – Social Policy	43	856
18	N4 – Accounting	44	5,846
19	G6 – Software Engineering	45	1,259
20	N6 – Human Resource Management	46	768

Source: UCAS Data, 2004.

by the average distances that applicants travel to accept places. These measures are partly related to the type of neighbourhoods from which a course typically attracts students, the scarcity of the course nationally and the typical location of the institutions in which the courses are delivered. The propensity for students from particular neighbourhood groups to attend Higher Education courses will be discussed later; however, the geographical location of those institutions offering a selection of those courses with high distance travel profiles are shown in Figure 4.8 and low distance profiles in Figure 4.9. Shown in these Figures are the two degree course groups with the highest average propensities to travel long distances which are Maritime Technology (J6) and Naval Architecture (H5), both of which are specialist course areas and being taught at a limited number of institutions (see Figure 4.8); and, degree courses with the lowest propensity to travel (Law by Topic – M2, Historical and Philosophical studies – V0, and Information Systems – G5) have a far wider distribution (see Figure 4.9).

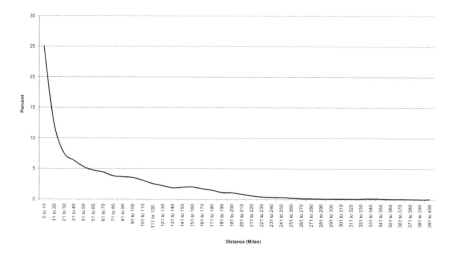

Figure 4.10   Distance decay chart of degree acceptances

Source: UCAS Data, 2004.

Using distance data it is possible to draw aggregate distance decay (Lösch, 1954) charts for institutions and neighbourhood Types, thus illustrating the effect that distance has upon accessibility to Higher Education, measured by the pattern of acceptances. Figure 4.10 shows the decay in acceptances in ten mile intervals from the institution. These aggregate data can be broken down into the distance decay between neighbourhood Types which show considerable variability. The difference in accessibility between the wealthiest ("Symbols of Success") and least wealthy ("Welfare Borderline") is shown in Figure 4.11.

Around 60% of all acceptances from "Welfare Borderline" (darkest line) neighbourhoods come from within 10 miles of an institution and compared with just 15% from "Symbols of Success" (lightest line) neighbourhoods. The mix of neighbourhood Types an institution attracts creates similar aggregate distance decay profiles. The institutions "M40" Manchester Metropolitan University and "M20" the University of Manchester, both of which are in the same city and within 1 mile of each other have the distance decay profiles which can be found in Figure 4.12. Manchester Metropolitan University (M40: Darkest Line) attracts 15% more of its applicants than Manchester University from within 10 miles of the institutions. Of further note is the latter peak visible in both universities at around 160 miles from the institution. This is the approximate distance from London, where because of good transport connections to Manchester and a high population, there is an increase in propensity to supply students. Furthermore, this pattern is more significant for the University of Manchester, where students will typically be attracted from more affluent neighbourhoods and as such have fewer restrictions on travel.

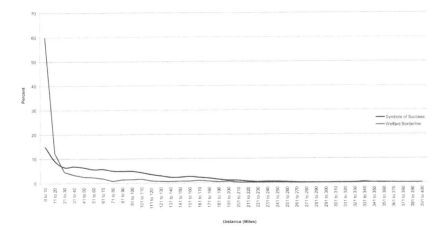

Figure 4.11   Distance decay chart of "Welfare Borderline" and "Symbols of Success" degree acceptances

Source: UCAS Data, 2004.

This section has shown how the location of an institution, the courses it has to offer and the tariff score of those accepted students (an indication of entry requirements) appear to influence the distance applicants from particular neighbourhood Types are willing or able to travel to accept an offer. These interacting spatial complexities affect both individual course and institutional market areas, and have implications for the targeting of particular student groups for purposes of marketing or widening participation.

Geodemographics and Social-Economics

The previous section has shown the apparent discontinuities in travel to accept places at Higher Education, both spatially and by neighbourhood Type. To further elaborate how variation in Higher Education access geographies are manifested, a series of charts based upon index scores are discussed in the following section. An index score can be used to show the overrepresentation of a target group by a discrete classification when compared to its proportions in a base population – as in Equation (4.1) where index scores *I* are calculated by comparing the proportion of a variable *v* within a target population *t* relative to a base population *b*.

Equation (4.2) –
$$I_v = \frac{\dfrac{t_v}{\sum\limits_{n=1}^{n} t_v}}{\dfrac{b_v}{\sum\limits_{n=1}^{n} b_v}} \times 100$$

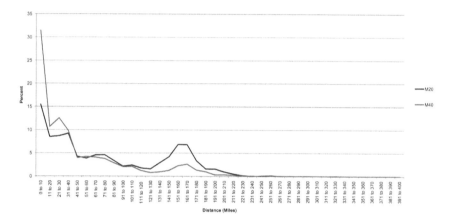

Figure 4.12   Distance decay chart for Manchester University – M20 and Manchester Metropolitan University – M40 degree acceptances

Source: UCAS Data, 2004.

These analyses illustrate how neighbourhood typologies can provide an effective tool to analyse socio-spatial differences in Higher Education data. An aggregate profile for 2004 home Higher Education acceptances can be created by comparing the proportion of all UK degree acceptances by neighbourhood Type with total adult population (see Figure 4.13). This shows how 2004 degree acceptance rates vary according to neighbourhood Types. There are large between Group differences, with "Symbols of Success" showing the greatest propensity to participate, and those from "Welfare Borderline", "Municipal Dependency" and "Blue Collar Enterprise" having the lowest. "Twilight Subsistence" and "Grey Perspectives" also have a low propensity to participate, although this is likely to be caused by the older age profile of residents of these areas. Some neighbourhood Groups exhibit high levels of within Group variability, with a mix of Type level participation rates above and below the average. Within the Group "Ties of Community", most Types have below average participation rates, with the exception of "South Asian Industry" and "Settled Minorities". There is an ethnic dimension to these

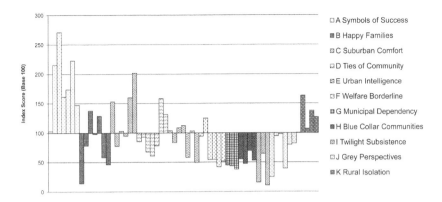

Figure 4.13 Degree acceptances by neighbourhood Type

Source: UCAS Data, 2004.

patterns, with both of these overrepresented Types having very high proportions of Asian ethnic minorities, and the other Types within this Group having a low proportion. The importance placed on Higher Education by people living within these neighbourhood Types could explain these differences. Similarly, within the Group "Welfare Borderline", the Type "Metro Multiculture" which defines neighbourhoods that typically have a high proportion of Asian residents correspondingly shows a higher propensity to participate in Higher Education.

In order illustrate a series of characteristics about the use of neighbourhood classifications within Higher Education, a profile for an alternative admissions system is presented in the following analysis. To qualify as a Primary or Secondary School Teacher, a qualification recognised by the Graduate Teacher Training Registry (GTTR) must be obtained. The GTTR administers applications to these courses through UCAS. The neighbourhood profile for 2004 acceptances to these GTTR courses is shown in Figure 4.14.

There is a high propensity for acceptances from the Group "Urban Intelligence", which are areas with people who are "young, single and mostly well-educated, and who are cosmopolitan in tastes and liberal in attitudes" (Experian, 2006). A further Type which stands out from the Group "Welfare Borderline" is "Bedsit Beneficiaries" whose inhabitants are generally "childless couples and singles renting in city centres from private or public landlords" (Experian, 2006). These areas have quite different social composition, yet both accommodate large numbers of students in shared accommodation. GTTR applications are for postgraduate study and are predominantly made in the final year of undergraduate study from a Higher Education term time address. Because of this, the registered address is unlikely to be that of the student's original (Pre-HE) domicile, thus explaining the high incidence of these Types.

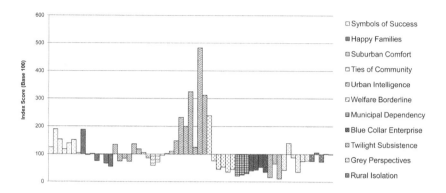

Figure 4.14   Propensity for 2004 GTTR acceptances by neighbourhood Type

Source: UCAS Data, 2004.

This also illustrates two further points about the use of geodemographics for Higher Education applications. When an application is made to UCAS, applicants are asked for both a "home" address and a "contact" address. The addresses used throughout this research are for "home" addresses, which are characteristic of the type of neighbourhood or background that a student has come from. There are inherent and undetectable errors with this process, as a student may apply for a new course of Higher Education while living away at university. It is possible that a student may specify the same "contact" and "home" address, as they perhaps view their current accommodation as their home. The GTTR profile also demonstrates a further issue, and one which makes targeting students during or post-undergraduate courses of Higher Education particularly challenging. The underlying theory of geodemographics is that a neighbourhood both defines, and is defined by the people who live within its area. Thus, as in common with socio-economic classification defined by the parental occupation (of applicants aged under 21), geodemographic classifications of Higher Education in the context of post-graduate education are predominantly defined not by the parental address, but rather the address/neighbourhood in which the student has lived. During the period of study student, behaviours may align more closely with those people living within the new neighbourhood containing the student accommodation. It is reasonable to anticipate change in student behaviour when living away from home, for example, with respect to the propensity to purchase beer. It is for this reason that the GTTR profile is defensible, as by the end of an undergraduate programme it is likely that actual student behaviour will be more homogeneous than the possibly more diverse range of neighbourhood Types defined according to parental home address.

Segmentation by Prior Attainment

Access to Higher Education requires an applicant to have prior attainment at A-Level or a nationally-recognised equivalent, in order to fulfil entry criteria that may stipulate a combination of grades and qualification types. Differences between students in terms of attainment therefore largely determine opportunities to access certain Higher Education courses or institutions and this, along with the demand for courses and subject profile, in turn affects an institution's aggregate profile of incoming students. For example, an applicant for a degree in Medicine would be expected to possess A-Levels in science subjects, and without these qualifications would not be successful in gaining access. These issues have been addressed in the literature by Reay et al. (2005:87) who argue that "choice in Higher Education is constrained by the predicted and actual grades achieved by the students". These patterns are further entrenched by requirements for specialism in prior qualification which are deemed essential by subject specialists in Higher Education institutions. These effects were recognised in the 2003 White Paper on Higher Education:

> The problem does not begin at age 16. Recent research suggests that a significant difference appears even before children have reached the age of two years. The attainment gap continues to widen through the phases of education, although the pace of increase slows down once children reach 7 years of age. Analysis suggests that at least three quarters of the 30 percentage point social gap in higher education participation can be attributed to differences in the level of attainment by the age of 16. Thirty per cent of children whose parents are in unskilled occupations achieve five or more good GCSEs, compared to 69 per cent of children whose parents are professional or managerial. (DfES, 2003:7)

Profiling attainment at Key Stage 4 (KS4) demonstrates the aggregate profile of qualified students who potentially could enrol in post-16 study, and as such, gain entry to university. GCSE grades can be converted into a point scheme which is used by the DCSF to calculate a number of measures on school attainment tables. The Qualifications and Curriculum Authority (QCA)[3] GCSE point scheme is shown in Table 4.3.

Table 4.3   QCA GCSE point scheme

Grade	A*	A	B	C	D	E	F	G
Points	58	52	46	40	34	28	22	16

---

3   www.qca.org.uk/.

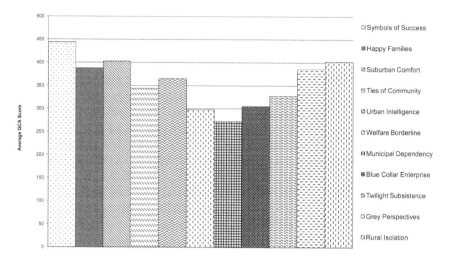

Figure 4.15   Average KS4 2006 GCSE points by Mosaic Groups

Source: DCSF.

Table 4.4   Index scores for 2006 KS4 GCSE grades by Mosaic Groups

Grade	A*	A	B	C	D	E	F	G
Symbols of Success	234	179	133	91	63	43	31	22
Happy Families	93	105	110	108	99	88	75	63
Suburban Comfort	126	125	118	104	89	75	61	51
Ties of Community	59	74	89	104	112	117	122	124
Urban Intelligence	139	119	105	93	93	89	89	95
Welfare Borderline	45	56	73	92	118	143	168	191
Municipal Dependency	17	32	53	88	125	168	213	262
Blue Collar Enterprise	36	51	70	97	123	147	164	167
Twilight Subsistence	58	69	85	102	114	124	129	133
Grey Perspectives	124	120	112	102	93	81	71	61
Rural Isolation	149	135	119	101	86	70	54	44

Source: DCSF.

Using this point system, an average score across neighbourhood Groups has been calculated (see Figure   4.15). There are clear differences between the neighbourhood Groups which can be examined further with reference to GCSE grade profiles (see Table 4.4). In general, those who are living in the most affluent areas achieve the highest grades. These data clearly show that aggregate educational attainment is both stratified and concentrated by neighbourhood Group at GCSE level, and as such, has a bearing on those students who will be qualified to pursue post-16 curricula.

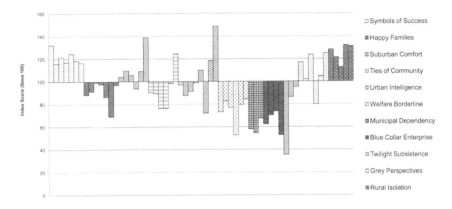

Figure 4.16   2006 Biology A-Level geodemographic profile

Source: DCSF.

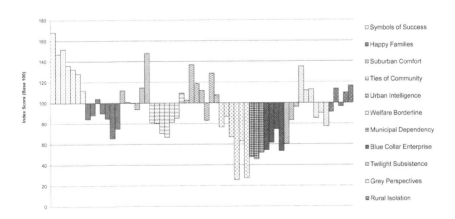

Figure 4.17   2006 Mathematics A-Level geodemographic profile

Source: DCSF.

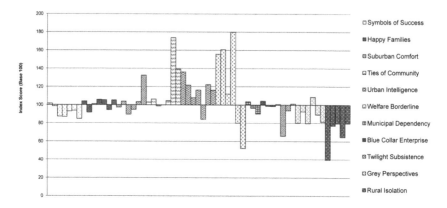

Figure 4.18   2006 Sociology A-Level geodemographic profile

Source: DCSF.

The previous analysis has shown how inequality exists between aggregate neighbourhood GCSE scores and the following section will show how similar patterns occur at A-Level. Before a student can attain grades in a subject at A-Level, they first must gain a place on a course at a school, college or other tertiary sector institution. These rates of access are variable which provides a first filter on those qualifying students who may be available to study courses of Higher Education. Thus, using the 2006 Key Stage 5 (KS5) PLASC data, Figures 4.16 to 4.19 present a series of geodemographic profiles for the areas in which students live who are studying particular A-Level subjects.

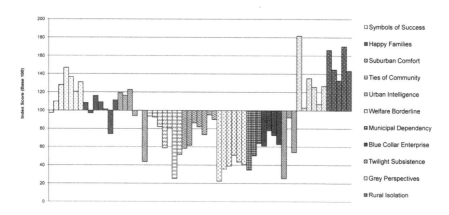

Figure 4.19   Geography A-Level geodemographic profile

Source: DCSF.

These variable rates of access could occur because certain neighbourhood Groups are inclined towards particular subjects, or that subject availability varies spatially. These effects filter those students who are available to apply for courses in Higher Education. Thus, for a Higher Education Geography department it becomes difficult on aggregate to attract students who have previously studied Geography at A-Level from less affluent backgrounds. However, these recruitment difficulties are shown to be concentrated if the attainment within these courses at A-Level is examined.

Table 4.5 shows the admissions criteria to Geography for the Russell Group as a sample of top research institutions, and those that often are criticised for not extending access. The entry requirements to study these courses are all very high, with all institutions requiring more than three B grades at A-Level and many also specifying that one of these must include Geography. This immediately limits

Table 4.5   2006 Geography degree entry requirements within Russell Group Institutions

University	Entry Requirement	Geography Required?
Cardiff University	300 pts*	Yes
Imperial College London	N/A	N/A
King's College London	BBB	Yes
London School of Economics and Political Science	BBB	No
Newcastle University	ABB	Yes
Queen's University Belfast	BBC–BCCb**	Yes
University College London	AABe–ABBe	Yes
University of Birmingham	ABB–BBB	Yes
University of Bristol	AAA–AAB	Yes (not for BSc)
University of Cambridge	AAA	No
University of Edinburgh	BBB	Yes
University of Glasgow	ABB	No
University of Leeds	ABB	Yes
University of Liverpool	320 pts	No
University of Manchester	AAB–ABB	No
University of Nottingham	AAA–AAB	Yes
University of Oxford	AAA–AAB	Yes
University of Sheffield	ABB–ABbb	Yes
University of Southampton	AAB–ABB	Yes
University of Warwick	N/A	N/A

Notes: * Minimum points required from qualifications with the volume and depth of A-Level or equivalent.

** Lowercase letters indicate an AS-Level.

those neighbourhoods from which the Higher Education institution can recruit, and as Figure 4.19 showed, not all neighbourhoods have equal propensity to study Geography.

Figure 4.20 demonstrates how this issue is compounded if attainment within Geography A-Levels is profiled. For example, in those institutions wishing to recruit "A" grade Geography A-Level students, there is bias towards students achieving these grades who live in more affluent areas. This makes it very difficult for Russell Group institutions to extend their offers into those areas which typically would not participate in Higher Education.

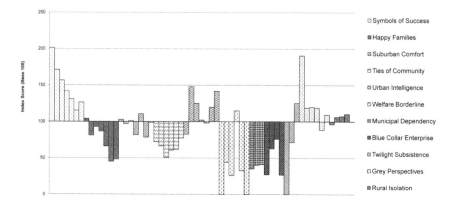

Figure 4.20   "A" at A-Level Geography in 2006

Source: DCSF.

The different aggregate preferences for students within neighbourhood Types to study subjects, and the resulting attainment stratification, mean that national and even institutional level targets to widen participation should address these differentials. For example, if a department runs courses which have lower entry requirements and typically attract students from widening participation neighbourhoods, this department may find it far easier to recruit these students than a high entry requirement department. From a national perspective, institutional widening participation targets should not be set without considering these prior educational settings, as the mix of Higher Education courses on offer, the quality/attainment of the local state school system and the courses they run could all affect the supply of qualified widening participation students.

Higher Education Course Choice and Neighbourhood Origins

Index scores are calculated for two JACS Course lines in Figure 4.21 and Figure 4.22 and show the propensity for participation from different Mosaic Types. For courses in subject line "A1 Pre-Clinical Medicine", Type index scores range between 120 and 184 within the Mosaic Group "Symbols of Success", indicating that students studying these courses are from more affluent neighbourhoods. In contrast to this profile the subject line "L5 Social Work" is shown in Figure 4.22.

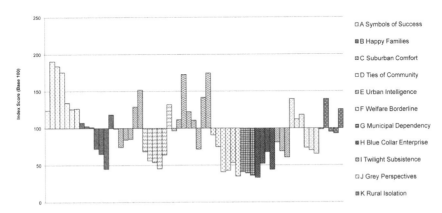

Figure 4.21   Participation to JACS Line Pre-clinical Medicine

Source: UCAS Data, 2004.

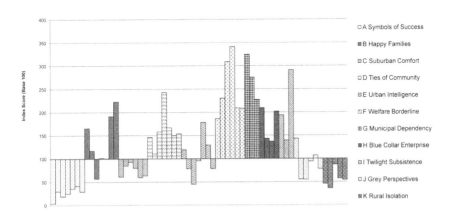

Figure 4.22   Participation to JACS Line Social Work

Source: UCAS Data, 2004.

In this subject line, neighbourhoods which are mainly urban and deprived are over represented. Those very wealthy Types making up the "Symbols of Success" Group are underrepresented within this subject of study. Any institution has a finite limit on the number of places it can offer applicants, restricted by the sum of the total capacity of those active courses across the institution. However, demand for these courses will be uneven, leaving some courses with surplus places and others with surplus demand. Within a constrained number of course places, geodemographic analysis could provide two useful tools to an institution to maximise the efficiency of the offer process. An applicant applying to an oversubscribed course could potentially be offered a place on a course that is in less demand, but informed by the information that there is a high probability that applicants from this neighbourhood Type would accept this offer. For example, courses in Nutrition may be oversubscribed, although surplus applicants might be redirected towards available places in Anthropology rather than rejecting them outright, as there is a high propensity for acceptances to both these subject lines from the neighbourhood group "Urban Intelligence". More strategically, widening participation bursaries could be distributed to those courses which have the greatest under representation of those neighbourhood Types typically underrepresented in Higher Education. If, for example, a course had 30 fewer students from these neighbourhood groups than would be expected, it may be that 30 bursaries should be provided to encourage these students to participate. However, provision of such bursaries does not guarantee that applicants will apply or indeed accept offers, and therefore the rate of return on bursaries available would have to be factored into the number offered, e.g. 50 may need to be offered to get a return of 30 acceptances. Furthermore, such initiatives could be described as attempts at social engineering, and any attempt to alter course profiles will reduce the probability of non targeted students being allocated or accepting places on these courses. A decision may therefore be taken to target only those courses that have surplus capacity. This raises the important issue of whether access to Higher Education can be "fairly" extended to underrepresented groups in institutions where capacity remains fixed and oversubscribed. One view would simply be that the best qualified students alone should be offered places, yet because of the inequalities in the level (and in some cases type) of incoming attainments from widening participation groups, this would always result in them being under-represented. A method of accounting for social inequalities in the offer making process is necessary in order to achieve a socially broader allocation of places across both over and under subscribed courses.

Demographic, Social and Economic Considerations

The index scores in Table 4.6 show the propensity for 2004 UK degree acceptances by conventional socio-economic groups across four age bands. In the youngest group, aged 20 and under, there is an above average participation rate for groups 1, 2, 4 and 5. It is interesting to note that group 3, "Intermediate Occupations", has only an average propensity to participate. Participation across groups 6 and 7 are at or below the average, which is as one would expect considering that these

Table 4.6   Index scores for socio-economic group by age

	20 and under	21 to 24	25 to 39	40 +
1. Higher managerial and professional occupations	117	28	29	39
2. Lower managerial and professional occupations	109	53	67	85
3. Intermediate occupations	98	96	123	116
4. Small employers and own account workers	113	49	47	48
5. Lower supervisory and technical occupations	114	45	38	44
6. Semi-routine occupations	90	144	145	121
7. Routine occupations	100	124	88	55
8. Unknown	71	233	207	196

Source: UCAS Data, 2004.

groups are less affluent. As discussed previously, this principal student age cohort is assigned to a socio-economic group based on the parental occupations, which not only requires the applicant to know what the job the parent does, but also for the short description of this job to be correctly coded into a classification. After the age of 21, the classification is based on the occupation of the applicant, and so the data are not directly comparable. This likely accounts for the decrease in the size of groups 1 and 2 across the older age cohorts. There is also increased propensity in the three older groups for the applicant to be coded as "unknown", thus indicating that there may be some systematic error in the conversion of occupation to socio-economic group. Overall, 20% of applicants are coded within this category and this undermines the usefulness of the classification.

Figure 4.23 shows the neighbourhood profile for those student acceptances aged over 19 years. This chart demonstrates how mature students more predominantly come from neighbourhoods which are less affluent when compared to young participants.

Although Higher Education is not restricted to certain age cohorts, the majority of people who accepted places to study degrees in 2004 were aged between 17–25, with a peak at aged 18 (see Figure 4.24). Thus the selection of base scores which are used to calculate neighbourhood participation rates affects the outcome significantly. Figure 4.13 showed the participation rate against a national population base, and by interpreting these statistics, we define Higher Education as available for all; that is, we should be equally worried about non participation across all age ranges. However, the majority of policy initiatives to extend participation to underrepresented groups focus on young participants, i.e. those students aged between 18–19 or under 21. In order to calculate a statistic which represents these policy target groups, a base aligned with the age range 18–19 was derived. One issue

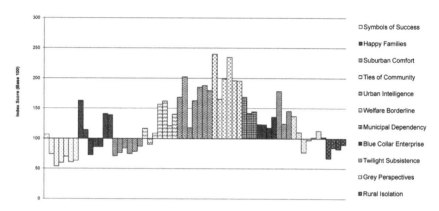

**Figure 4.23   2004 degree acceptance profile excluding 19 year olds
or younger by Mosaic Type**

Source: UCAS Data, 2004.

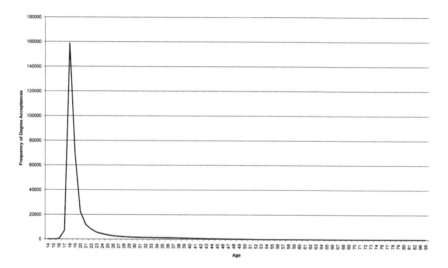

Figure 4.24   Age distribution of 2004 degree acceptances

Source: UCAS, 2004.

with postcode level geodemographic indicators is that national coverage population
data are not disseminated at this scale across different age ranges. Thus, a method
of disaggregating Census data at OA level into unit postcodes was devised.

The Office for National Statistics supplies a file which contains frequencies
for population, households and gender for each unit postcode for England and
Wales derived from the 2001 Census. The 2001 Census 18–19 year old population

available at OA level was assigned between unit postcodes within these areas using proportions created from the total population. This is derived using Equation (4.3) where p_i is the absolute population within a unit postcode, j; the population of 18–19 year olds within Output Area y.

$$\text{Equation (4.3)} - \qquad j_y \times \frac{p_{iy}}{\sum\limits_{i=1}^{n} p_{iy}}$$

This algorithm was implemented in SAS[4] Macro language. The macro iterated through a list of all OA, and using a mixture of SQL and SAS statements, created a series of temporary tables in which calculations occurred. At the end of the application, all temporary tables were amalgamated to create a new file of postcodes and their estimated frequencies of 18–19 year olds. These scores were deliberately not rounded, so a postcode could contain a proportion of a person. The postcodes were later aggregated into the 61 Mosaic Types, and so unit postcode level rounding would have introduced an unnecessary margin of error. In Figure 4.25 the estimated 18–19 year old base is compared with the same age range for those accepting degree places in England and Wales.

There is clear overrepresentation of Types falling within the Groups "Symbols of Success", "Happy Families", "Rural Isolation" and "Suburban Comfort", with the exceptions of Types "In Military Quarters" and "Burdened Optimists".

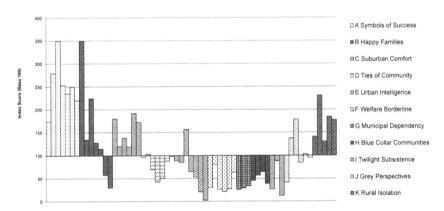

Figure 4.25   Participation rates by Mosaic Groups and Types in the 18–19 age cohort

Source: ONS and UCAS, 2004.

---

4   www.sas.com.

The former of these Types contains very small counts and the index score may be anomalous. People living within "Burdened Optimists" areas are reasonably affluent, living in mortgaged property, and not having particularly high levels of education. The appearance of this Type in a typically high participation Group indicates a need for bespoke geodemographic classifications that better account for these within Group differences. A further Group demonstrating a heterogeneous Type distribution is "Grey Perspectives". The first Type within this group has a very small base and target score, indicating that this index is probably unreliable. The second and third types within this group are called "Child Free Serenity" and "High Spending Elders". Both Types reside in affluent areas with high levels of educated residents. Students supplied from these neighbourhoods may therefore be the last in a line of siblings to attend university.

The concept of widening participation or extending access was introduced in Chapter 2 and the following analysis will evaluate this concept in quantitative terms through examination of current national indicators and benchmarks for widening participation. Aggregate indicators for institutions provide a method of assessing performance on a range of directly and indirectly recorded criteria. A number of quantitative indicators have been used by the funding councils to allocate widening participation funding and to assess widening participation performance. These have included:

- Higher Education Funding Council for England (HEFCE) POLAR
- National Statistics Socio-Economic Classification (NS-SEC)
- Proportion of state school students
- Geodemographics (Super Profiles)

Funding is given to institutions on the basis of their performance in widening access to young students (<21 yrs) from disadvantaged backgrounds. The allocation of this money is currently calculated using a classification called POLAR. Five quintile bands were used to allocate £92.3m of funding in 2006/07 (HESA, 2006) and only those students in the lowest two participation bands qualify for an institution to receive funding. A key geographical problem with POLAR measurements of ward level participation rates that is not widely recognised is the heterogeneity which may occur within the ward. On average in England a ward contains 5,500 people,[5] and by assigning a single classification of participation rate assumes within ward homogeneity. Particularly in dense urban areas where the scale of socio-economic differentiation often occurs at the level of the individual (Harris and Longley, 2005), these broad geographical aggregations fail adequately to measure micro level differentiation of population characteristics.

Although POLAR is used in the assignment of widening participation funding, it is not currently used in Higher Education performance indicators. Neighbourhood participation rates are currently measured using Super Profiles

---

5   www.statistics.gov.uk/geography/electoral_wards.asp.

which is a geodemographic classification based on the 1991 Census of Population. These are outdated and have been superseded by more modern classifications such as the 2001 versions of Mosaic from Experian and ACORN from CACI.

Other national measures of widening participation performance used in the performance indicators include NS-SEC and the proportion of state school students, and although they may demonstrate spatial autocorrelation, they are attributed using individual characteristics and not geographical locations. As discussed previously, the former of these classifications is based on the parental occupation of applicants under the age of 21, and when over the age of 21, their personal occupation. The proportion of state school students is derived using the UCAS classification of state and independent schools.

In order to better understand how neighbourhood classifications relate to both Higher Education widening participation performance indicators and funding mechanisms, a cross tabulation between Mosaic Types, POLAR, NS-SEC and the proportion of state schools was created.

Figure 4.26 and Figure 4.27 illustrates how the lowest two quintiles of the POLAR classification cross tabulate with Mosaic. As with previous index scores, those bars above 100 indicate neighbourhood Types which are overrepresented in Higher Education, and include the most deprived Groups of "Welfare Borderline" and "Municipal Dependency". However, there are apparent anomalies in these results, including the overrepresentation of a number of Types within the Group "Happy Families" – which based on their socio-economic characteristics, would have few restrictions on their propensity to participate in Higher Education. These results are created from all 2004 degree acceptances and illustrates a potential systematic problem within the POLAR classification.

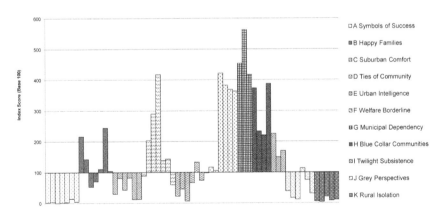

Figure 4.26   Very low (<16%) POLAR participation groups by Mosaic Groups and Types

Source: HEFCE.

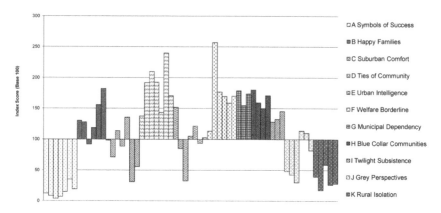

**Figure 4.27   Low (16–24%) POLAR participation groups by Mosaic
                Groups and Types**

Source: HEFCE.

The results from the cross tabulation for NS-SEC are shown in Figure 4.28, Figure
4.29, Figure 4.30 and Figure 4.31. Figure 4.28 shows an overrepresentation of the
Group "Rural Isolation" and demonstrates how univariate classifications based
on occupation interact with the geographic arrangement of typical industry types.
These neighbourhood Types might be described as affluent, and therefore should
not necessarily be considered as areas where widening participation students might
live. It could also indicate an inadequacy in Mosaic at capturing areas of rural

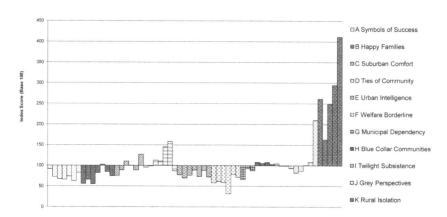

**Figure 4.28   The propensity for degree acceptances in Group 4: Small
                 employers and own account workers by Mosaic Groups and Types**

Source: UCAS Data, 2004.

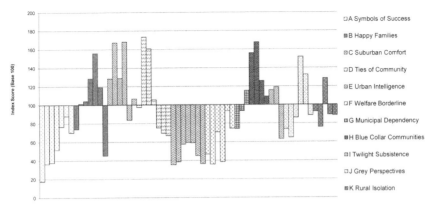

**Figure 4.29   The propensity for degree acceptances in Group 5:
Lower supervisory and technical occupations by Mosaic
Groups and Types**

Source: UCAS Data, 2004.

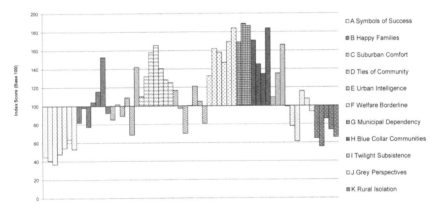

**Figure 4.30   The propensity for degree acceptances in Group 6:
Semi-routine occupations by Mosaic Groups and Types**

Source: UCAS Data, 2004.

poverty accurately. Similarly, Figure 4.29 shows an over representation of more affluent neighbourhood Types, with the exception of "Blue Collar Communities". Figure 4.30 and Figure 4.31 show these students to have a higher propensity to live in less affluent neighbourhoods, and more in line with those areas where it might be considered desirable to target widening participation funding. From the results of this geodemographic analysis, there is compelling evidence to suggest that NS-SEC performance indicators might be better created simply from the lowest two groups, as these best represent low participation neighbourhoods.

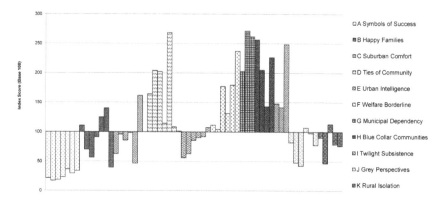

**Figure 4.31   The propensity for degree acceptances in Group 7:
Routine occupations by Mosaic Groups and Types**

Source: UCAS Data, 2004.

Figure 4.32 shows the propensity for students previously educated in state schools to accept degree places, broken down by Mosaic Type. Because the majority of students study within state schools, only those students from neighbourhoods which are particularly affluent, and as such have increased ability to pay school fees show a low propensity for state school attendance. The differentiation between neighbourhood Types for this particular indicator is therefore quite low.

An unadjusted participation rate for 2004 degree acceptances was presented earlier. This was created by comparing the proportions of total UK degree acceptances with the total UK population by Mosaic Types. The patterns of neighbourhood

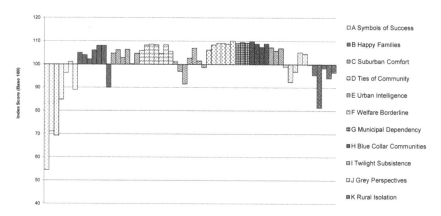

**Figure 4.32   The propensity for state school student degree acceptances
by Mosaic Groups and Types**

Source: UCAS Data, 2004.

disadvantage are different from the national indicators and provide evidence that they may not identify low participation neighbourhoods correctly. In order to quantify how the national widening participation indicators discussed earlier (including <16%POLAR, 16–24% POLAR, NS-SEC 4, NS-SEC 5, NS-SEC 6 and NS-SEC 7 and State Schools) relate to the actual participation rates at neighbourhood level, the index ranking by the 61 Mosaic Types for each of these national measures was compared to the actual ranking of participation. Additionally, Experian provide a ranking of Mosaic Types by "Wealth" which is based on aggregated consumer income data. The rankings for these various measures are compiled in Table 4.7.

A Spearman's Rank Correlation (r_s) was calculated from these data to assess overall correlation between the actual 18–19 year old participation profile ("HE" Column in Table 4.7) and the national classifications where d_i are the differences between ranks x and y, and n are the frequency of types (see Equation 4.4). Included in the table of ranks (Table 4.7) is a wealth rank. This ranking is provided by Experian in the supplementary material accompanying Mosaic, however, in this instance the inverse ranking was used, and as such is a measure of low wealth and therefore should be positive correlated with indicators that show increased propensity in less affluent areas.

Equation (4.4) –
$$r_s = 1 - \frac{6 \times \sum d_i^2}{n \times (n^2 - 1)}$$

Table 4.8 shows the results of Spearman's Rank Correlation.

Of these indicators, the highest correlation with actual participation rates is tied between POLAR <16% and NS-SEC 6 with a score of 0.77. One would expect a high correlation with the lowest POLAR quintile as it is the only national indicator that directly measures low participation. It is also reassuring to find quite a strong correlation between the POLAR 16–24% class and neighbourhood level participation. Oddly, NS-SEC 7, which is less "affluent" than NS-SEC 6, has a lower correlation with actual neighbourhood participation rates, indicating an issue with this particular indicator. As might be expected given the index profiles presented in Figure 4.28 and Figure 4.29, NS-SEC 4 and 5 shows very low correlation with participation rates, with correlations of 0.01 and -0.03 respectively. State school attendance shows quite a strong correlation of 0.64, although this is one of the lower correlations. Neighbourhood wealth has a correlation of 0.83 which is higher than any of the national benchmarks, indicating a very strong relationship between financial means and participation in Higher Education. The interesting conclusion from this analysis is that the indicators used in Higher Education as benchmarks do not correlate highly with neighbourhood participation rates, providing evidence that these measures may be equally ineffective in performance monitoring and funding assignment. Furthermore, the best performing indicator is one which is not based on public sector data sources, suggesting the need to access indicators of wealth, perhaps through public/private sector data sharing partnerships.

Table 4.7 National indicators of low participation ranked by Mosaic Types

	HE	POLAR <16%	POLAR 16–24%	NS-SEC 4	NS-SEC 5	NS-SEC 6	NS-SEC 7	State School	Wealth
A01 Global Connections	15	59	58	28	62	59	60	61	47
A02 Cultural Leadership	3	57	59	49	59	61	62	59	56
A03 Corporate Chieftains	2	61	61	52	57	62	61	60	61
A04 Golden Empty Nesters	4	60	60	55	50	58	59	57	60
A05 Provincial Privilege	6	58	57	46	34	56	56	50	58
A06 High Technologists	5	47	49	56	31	53	58	41	59
A07 Semi-Rural Seclusion	9	55	55	37	42	57	57	56	57
B08 Just Moving In	1	16	26	60	40	44	25	30	26
B09 Fledgling Nurseries	21	21	28	53	23	37	45	35	48
B10 Upscale New Owners	8	38	39	61	22	47	48	38	55
B11 Families Making Good	23	35	30	40	11	33	36	25	44
B12 Middle Rung Families	26	26	17	19	7	27	21	16	43
B13 Burdened Optimists	42	12	6	36	15	13	19	17	25
B14 In Military Quarters	52	28	37	45	53	40	55	55	36
C15 Close to Retirement	12	43	43	44	13	43	47	31	54
C16 Conservative Values	24	32	32	29	4	34	34	23	50
C17 Small Time Business	19	40	40	11	10	41	40	36	46
C18 Sprawling Subtopia	25	31	24	20	2	30	32	22	45
C19 Original Suburbs	10	49	51	31	33	50	52	42	52
C20 Asian Enterprise	16	48	45	9	20	16	15	32	27
D21 Respectable Rows	30	30	23	25	24	28	31	26	35

Table 4.7 continued   National indicators of low participation ranked by Mosaic Types

	HE	POLAR <16%	POLAR 16-24%	NS-SEC 4	NS-SEC 5	NS-SEC 6	NS-SEC 7	State School	Wealth
D22 Affluent Blue Collar	27	17	5	24	1	21	14	15	34
D23 Industrial Grit	38	11	3	10	5	12	9	11	24
D24 Coronation Street	47	4	4	12	21	9	11	14	13
D25 Town Centre Refuge	45	22	22	8	36	17	22	33	18
D26 South Asian Industry	34	20	2	7	43	22	2	12	8
D27 Settled Minorities	29	37	12	34	44	23	26	28	21
E28 Counter Cultural Mix	32	44	19	42	61	25	30	40	20
E29 City Adventurers	35	39	41	51	55	38	49	47	39
E30 New Urban Colonists	17	53	50	43	49	49	46	54	49
E31 Caring Professionals	39	36	35	30	47	35	39	37	31
E32 Dinky Developments	44	23	29	47	48	24	37	20	30
E33 Town Gown Transition	59	34	38	32	52	32	35	39	23
E34 University Challenge	61	29	36	48	58	45	27	45	15
F35 Bedsit Beneficiaries	51	24	33	59	51	20	24	24	14
F36 Metro Multiculture	37	27	1	57	60	10	29	13	7
F37 Upper Floor Families	56	3	9	58	41	11	13	6	6
F38 Tower Block Living	58	7	14	62	56	14	20	7	1
F39 Dignified Dependency	54	9	16	41	26	6	12	10	5
F40 Sharing a Staircase	41	10	13	50	37	3	6	1	2
G41 Families on Benefits	57	2	8	54	38	7	10	5	3
G42 Low Horizons	53	1	18	27	25	1	1	3	4

Table 4.7 continued   National indicators of low participation ranked by Mosaic Types

	HE	POLAR <16%	POLAR 16–24%	NS-SEC 4	NS-SEC 5	NS-SEC 6	NS-SEC 7	State School	Wealth
G43 Ex-industrial Legacy	50	5	10	33	18	2	3	4	10
H44 Rustbelt Resilience	46	8	7	13	6	5	4	2	12
H45 Older Right to Buy	43	13	15	16	3	15	8	9	22
H46 White Van Culture	40	15	20	14	14	19	17	18	19
H47 New Town Materialism	49	6	11	18	19	4	7	8	11
I48 Old People in Flats	55	14	27	17	17	29	16	19	9
I49 Low Income Elderly	33	19	25	22	16	18	18	27	16
I50 Cared for Pensioners	60	18	21	23	46	8	5	21	17
J51 Sepia Memories	48	41	46	26	39	36	41	44	32
J52 Childfree Serenity	20	46	47	39	45	46	50	53	42
J53 High Spending Elders	13	50	52	35	32	54	53	48	53
J54 Bungalow Retirement	36	25	31	21	8	26	28	29	41
J55 Small Town Seniors	28	33	34	15	9	31	33	34	29
J56 Tourist Attendants	31	42	42	5	30	39	43	43	28
K57 Summer Playgrounds	18	54	48	3	27	52	38	51	38
K58 Greenbelt Guardians	7	56	56	6	35	55	51	58	51
K59 Parochial Villagers	22	45	44	4	12	42	23	46	37
K60 Pastoral Symphony	11	52	54	2	28	48	42	52	40
K61 Upland Hill Farmers	14	51	53	1	29	51	44	49	33

Source: UCAS, 2004 and Experian.

Table 4.8   Spearman's rank results (low actual participation)

Classification	Spearman's Rank
POLAR <16%	0.77
POLAR 16–24%	0.66
NS-SEC 4	0.01
NS-SEC 5	-0.03
NS-SEC 6	0.77
NS-SEC 7	0.69
State School	0.64
Wealth	0.83

Conclusion

This chapter has presented an exploratory data analysis predominantly using index scores and has shown how access to Higher Education is both spatially and socially heterogeneous. The distance that accepted applicants travel to start their degrees has been shown to be stratified spatially by both course and institution. The fewer locations that a course is offered, the further (on average) students travel to accept these places as they are offered at a restricted selection of potential institutions. Institutions with high entry qualifications were shown to attract students from a broader area. When average tariff scores were examined by neighbourhood Types, higher attainment was recorded in students from more affluent neighbourhoods, thus having fewer restrictions on travel, both socially and economically. A model was created which estimated the 18–19 year old base population in England and Wales at unit postcode level and was compared to the same age range accepting Higher Education places in 2004. This was aggregated to create a measure of neighbourhood participation. This ranking of neighbourhood level young participation was compared to those rankings created by the classification used by HEFCE and HESA to allocate widening participation funding and assess benchmarked performance. Variable performance was found, with high correlation with the lowest POLAR and NS-SEC 6. Some issues requiring further investigation were the lower correlation with NS-SEC 7 and 4. The highest correlation was recorded using the Experian "wealth" ranking.

Chapter 5
Creating Open Source Geodemographics

The Public Sector and Geodemographic Classifications

There are a variety of commercial geodemographic classifications which can be used to inform spatial decision making in Higher Education, however, none have been specifically designed for this purpose. Although geodemographic classifications originated in the public sector (as a method of targeting deprived areas – see Webber, 1977; Webber, 1978; and Webber and Craig, 1978) they were subsequently augmented with private sector data for commercial applications such as customer segmentation and direct mailing (Harris et al., 2005). Public sector data should be made more readily available, although, even with its current limited dissemination, no current commercial vendor incorporates public sector transactional data at clustering unit level into commercial classifications. Experian were the first commercial vendor to provide a public sector version of their commercial classification, however, this has only been attained superficially at the level of "pen portraits" which provide additional descriptive material of the clusters. Although these classifications may be promoted as bespoke solutions for the public sector, they do not address a number of key concerns, including whether it is appropriate for a general purpose classification describing consumption of goods and services supplied by the private sector, should be applicable to public goods that are consumed collectively. Furthermore, for public sector use of geodemographic classifications, robustness in terms of social equality is acutely important as the misspecification of areas could have far reaching negative impacts. For example, in an advertising campaign targeting educational funding opportunities, residents of an incorrectly prioritised area may lose real life chances by not receiving appropriate information, despite being stakeholders in the educational and taxation systems. Because of these issues, greater transparency of classification procedures is required by public sector end users, including a greater level of methodological detail than has hitherto been provided by commercial vendors. The requirement for an open methodology is not easily achievable by commercial companies as the release of detailed information on how classifications are constructed could be perceived as undermining their competitive advantage. To date, details on classification methodology have only been made available at rather general levels, including detail of which clustering algorithms have been used (Harris et al., 2005) or the sources and broad mix of input data (Experian, 2006). It is in this context that this chapter explores how a bespoke geodemographic classification can be created by combining public domain and sector-specific data, using explicitly specified techniques and tools. This analysis extends a pre-existing classification from the Office for National Statistics called the Output Area Classification.

Clustering Methods and Global Optimisation

Numerous clustering algorithms are available to create groupings from large multidimensional datasets and a thorough review can be found in Everitt (1974). These methods all aim to create homogeneous groups from a multidimensional data matrix. In relation to geodemographics it has been argued that the notion of discrete clusters can be misleading (Voas and Williamson, 2001), and a more appropriate conceptualisation should be of overlapping clouds (Longley and Goodchild, 2007). This draws on a main dichotomy in clustering methods, between those that consider attribute space as a continuum, and those that seek to partition space into discrete groups. This fuzzy logic (Zadeh, 1965) opposes the Aristotelian-logic paradigm of science (Openshaw, 1998), proposing instead that "many, if not most, statements are indeterminate" (Kosko, 1993). The choice of clustering methodology should depend on the structure of the input data used to form the classification, and Kendal (1966) draws a distinction in cluster analysis between "classification" and "dissection". Classification implies that there are distinctive groups, and that "objects within groups shall be dissimilar from objects in other groups" (Gordon, 1981:5). By contrast, dissection is the process of "dividing homogeneous data into different parts" (Everitt, 1993). These two different methods are illustrated in Figure 5.1.

In socio-economic clustering applications such as geodemographics, it is proposed that dissection is a more accurate representation of the function of the clustering algorithm. The input data for the cluster analysis has high dimensionality across multiple attributes, and the range of values associated with these variables can vary widely across different spatial locations. On this basis, the logic of a fuzzy classification (dissection) is reinforced, as different

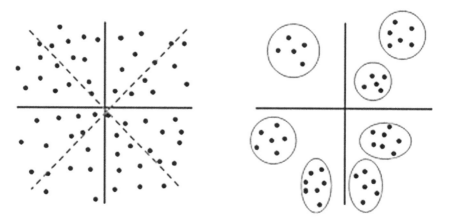

Figure 5.1   Dissection (left) and classification (right) cluster analysis

Source: Adapted from Gordon, 1980:4.

discrete dissections of the data could result in multiple classification schemes. With a fuzzy classification the blurring of boundaries between clusters is accommodated through a membership statistic which shows how well the data points fit within particular groupings. Although some have argued that a discrete classification "can be a gross over simplification of the structure in the data set" (Gordon, 1981:58), the prevalence of these classification types in commercial geodemographic applications has remained unanimous. A plausible explanation for this fact could be that end user communication is easier for discrete rather than for fuzzy classifications. For example, it is easier to describe that someone resides in neighbourhood group *a* or *b*, rather than a mix of multiple groups, and can be particularly confusing when ambiguous assignments are supported by supplementary descriptive materials.

Before running a cluster analysis, data must be standardised to potentially reduce the effect of outliers, measure the data on the same scale, and ensure that all variables have the same basic weighting. Romesburg (1984) discusses how standardisation prior to clustering prevents the units used for attribute measurement from affecting the similarities between objects, and therefore allowing unbiased contribution by each of the variables. There are many ways in which the data could be standardised and these are evaluated by Vickers et al. (2005), who found that in the production of the National Statistics Output Area Classification (OAC), the range standardisation method (Wallace et al., 1996) performed most effectively at reducing outlier effects in a Census based Output Area (OA) level classification. This method standardises all data between the scores 1 and 0. The OAC classification is built using the *k*-means algorithm which partitions an *R* multidimensional data matrix into *k* clusters or groups based on a local optimisation criteria. For a full evaluation of why this algorithm is most suited to clustering applications at Output Area geography, see Vickers (2005).

The *k*-means method (MacQueen, 1967) is an iterative relocation algorithm, which assigns the OA data points into *k* clusters based on a standardised Euclidean minimum distance metric. The algorithm sets the initial location of the cluster centroids As a set of random *k* locations within the *R* dimensional space. The distance of the data points to each cluster centroid is then calculated, and each data point is provisionally assigned to its nearest cluster centre. A clustering criterion statistic is then applied to measure the homogeneity within these temporary cluster allocations. Pythagoras' Theorem is applied in a R_n dimensional space (Gordon, 1981) where a dataset vector comprises of *n* variables (dimensions) weighted by an associated population size. At each iteration of the model, the population weighted distance between the data points and the cluster centroids are re-calculated using Equation (5.1).

Equation (5.1) –
$$d_{iQ} = \sum_{p=1}^{n} (x_{iR} - x_{QR})^2$$

This equation is applied to all clusters within a single iteration of the model, i.e. k_n clusters, and it is the sum of these results which form f Equation (5.2), a model objective function assessing the overall model performance or within cluster homogeneity.

Equation (5.2) –
$$f = \sum_{k=1}^{n} (d_{iQ})_k$$

After the first iteration of the model where initial cluster centroids (seeds) are randomly placed and all data points are temporarily assigned to their nearest seed, the k-means algorithm attempts to find a local optimum through an objective function that reallocates data points iteratively from their initial assignments. Each data point is considered for reallocation to other clusters, and after each test, the model objective function is recalculated. If the outcome is larger, i.e. a less homogeneous cluster, no further reallocation of data points takes place. Where reassignment of data points does occur, the cluster centroid values for the gaining and losing clusters are recalculated. The maximum number of iterations for this optimisation process can be specified by the user. However, with current computational power, it is possible to leave the models running until the iteration process converges, i.e. further reassignments of data points does not improve the sum of squares statistic. Everitt (1974:26) astutely observes that "there is no way of knowing whether or not the maximum of the criterion has been reached". This is because in a single k-means model there are multiple local optima, since the random placement of the initial cluster seed centroid means that there are multiple possible locally optimised models. This can be illustrated when two separate models are run to convergence where $k = 9$ on a two variable dataset extracted from the OAC input data (see Figure 5.2 and Figure 5.3).

Ignoring the arbitrarily assigned cluster names, these graphs show how the path of the cluster centroid can converge upon entirely different locations, depending upon the random initial seed location. Furthermore, as each iteration of the model reallocates data points to cluster centroids, "making the "best" decision at each particular step does not necessarily lead to an optimal solution overall" (Harris et al., 2005). The most effective partitioning of the input data in a cluster model is globally optimised, although in reality this is not obtainable, given that there is no benchmark of global model performance for an individual data set. However, with sufficient computational power, a globally optimised local model can be obtained by running k-means multiple times to convergence, comparing the results from each cluster analysis and saving the best performing classification. Figure 5.4 shows the results from a k=9 model which was run with a random seed allocation 150 times, and for each model an R-squared statistic was generated to estimate the quality of the model discrimination. This graph highlights the variance in overall model performance through selection of different initial seeds.

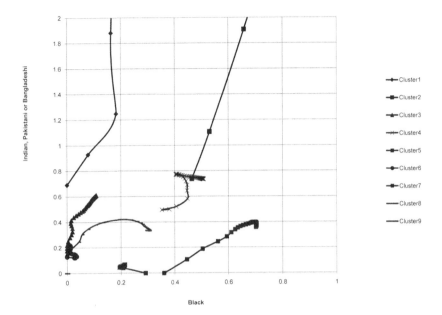

Figure 5.2   Cluster mean centroid paths – run 1

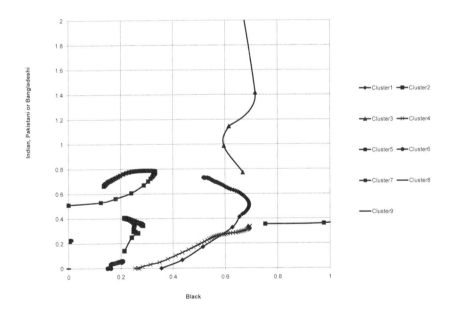

Figure 5.3   Cluster mean centroid paths – run 2

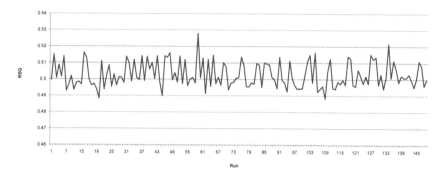

Figure 5.4   *R*-Squared results from 9 cluster model

A further potential problem with the *k*-means algorithm arises in selecting an appropriate number of clusters. The earlier discussion on the differences between dissection and classification demonstrates that there may not be an appropriate *k*, and that *k* is arbitrarily assigned to "slice" the data into reasonable, but not an ideal, set of groups. Vickers (2007) in the construction of OAC quotes geodemographic practitioner judgements on appropriate cluster frequency, based on their knowledge of what has been successful in commercial products, selecting three hierarchies of 7, 21 and 52, each successively partitioning the OA into smaller aggregations. A further method initially demonstrated by Debenham (2001) is to calculate the average distance between the data points and their assigned cluster centroid at model convergence for a range of different values of *k*. A judgement can then be made on an appropriate number of clusters, weighing up the costs and benefits between a frequency of clusters which can be easily described and interpreted by end users, and also a number which yields reasonably homogeneous characteristics.

   In addition to the decision of how many clusters are required in the final model, is the choice between two potential methods which could be used to construct an educationally weighted classification based upon OAC:

1. Re-cluster a new classification using raw data from OA level upwards including the original OA data and educational data aggregated by OA.
2. Use the existing OAC classification to create a further tier in the hierarchy by re-clustering each Sub Group into a number of "Micro Groups", and then by adding sector data, re-cluster these into a bespoke educational classification.

There are a number of problems associated with the former option. Firstly, the classification would require to be re-created from base principles and therefore have no likely resemblance to the existing Output Area Classification. This could result in difficulty when communicating these groups to end users of the existing classification. Secondly, unlike the variables used in the OAC classification taken from decennial census data, education data have a restricted socio-economic, and uneven geographical coverage, resulting in the increased prevalence of outlier values based on small base counts. These effects could be reduced though standardisation and careful population weighting, but they might still cause undesirable noise in the resulting classification. The second option potentially resolves these issues by providing a link to the existing classification and also a method by which small base count effects may be minimised because the clustering input are segments, i.e. aggregations of OAs containing similar socio-demographic characteristics. Therefore, the first stage in the analysis was to create a new tier in the OAC classification. A similar product called Mosaic Segments (Experian, Nottingham, UK) provides a 243 cluster solution which is a finer disaggregation of its 61 Mosaic Types. Previous work using this classification (Singleton and Farr, 2004) has demonstrated that this level of aggregation is effective for re-clustering Education data.

The input data used to create this new finer level classification consisted of a series of standardised 2001 Census variables at Output Area level, and was the same data used to construct OAC. This UK dataset was split into 52 separate groupings of Output Areas, based on their assignments in the OAC Sub Group classification. These 52 datasets were separately re-clustered using the *k*-means algorithm implemented in the SAS statistical software.[1]

As discussed earlier in this chapter, a method to derive a globally optimised local model for a single dataset is to run the *k*-means cluster analysis to convergence with multiple initial random seed locations, each time comparing the performance of the final classification using optimisation criteria such as the *R*-squared statistic. In order to avoid the problem of poor initial seed selection which was discussed earlier, each of the 52 datasets had the *k*-means algorithm repeatedly run 10,000 times and the best results extracted.

Following the methodology by which the hierarchies of the OAC were created, each of the 52 Sub Groups were divided into a further level referred to as "Micro Groups". The aim was to create a classification with a similar frequency of clusters to commercial "segment level" classifications. The initial frequency of divisions was based on the distribution of 18–19 year olds within the OAC Sub Groups. The percentage distribution of 18–19 year olds by OAC Sub Groups can be seen in Figure 5.5.

Dividing all Sub Groups evenly would create some Micro Groups with very small aggregations of the target population, and as such create outliers. Re-clustering Micro Groups with outliers could result in a bespoke classification consisting of particularly large or small groups which are not ideally placed to discriminate between different groups within the population. The population distribution chart in Figure 5.5 was used to estimate the initial frequency of the divisions required to create the Micro Group classification. For the two Sub Groups with the lowest population counts, it was decided that no division was required, effectively

---

1 SAS is a commercially available statistical software: www.sas.com.

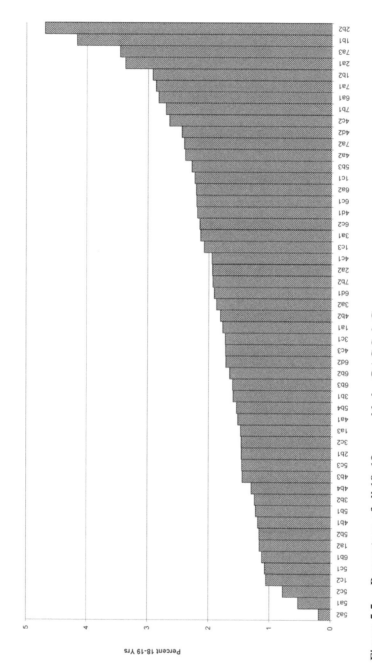

Figure 5.5   Percentage of all 18–19 year olds by OAC Sub Groups

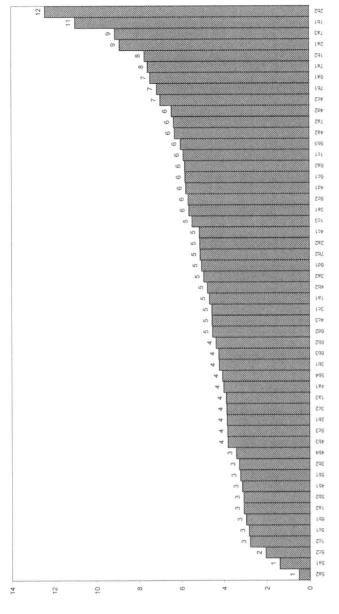

Figure 5.6   Frequency of divisions by OAC Sub Groups

re-using the OAC Sub Group classification. Under this constraint, the initial aim was to create 264 clusters in total, using those division frequencies shown above the bars in Figure 5.6. The division counts are created by apportioning the total Micro Groups within each Sub Group by the percentage of the 18–19 year old population present. The actual figures are represented using the bars, and the numbers above are the rounded figures used as the initial k values in the cluster analysis.

The classification created by these cluster analyses had a number of very small Micro Group clusters. In order to minimise these small clusters, and so maintain population uniformity within each Sub Group, a number of lower k values were reassessed to create more uniformly distributed clusters. The exact frequency of k which resulted in the most homogeneous division are shown in Table 5.1, however, numerous alternate values were tried. The final classification therefore divided all Output Areas into 176 Micro Groups and forms the base onto which the education data are appended.

Table 5.1   Re-assigned k values

Sub Group	Previous k	New k	Sub Group	Previous k	New k
1b1	11	4	4b3	4	1
1c3	5	3	4b4	3	2
2b2	12	5	4c2	7	2
2a1	9	3	4c3	5	4
2a2	5	3	4d1	6	2
2b1	4	3	4d2	6	3
2b2	5	2	5b2	3	2
3a1	6	1	6a1	7	5
3a2	5	2	6a2	6	2
3b2	3	2	6b1	3	1
3c1	5	4	6b1	3	1
3c2	4	2	7a2	6	4
4a1	4	1	7a3	9	5
4a2	6	1	7b1	7	5
4b1	3	2	7b2	5	4

Higher Education Variable Choice and Evaluation

The HESA data were not geographically referenced, so the students were geocoded using their home address unit postcode and linked to Output Areas through the National Statistics Postcode Directory [2] (NSPD). In order to create the Higher Education input variables, a series of binary (1,0) scores were calculated using

---

2   www.statistics.gov.uk/geography/nspd.asp.

various categorical attributes contained within the database. Using the OAC Micro-Group classification as an aggregating field, these binary values were summed along with the average scores from a series of continuous variables. The count variables were converted to index scores derived from the total distribution of students recorded in the HESA dataset, with the exception of the young participation variable, which used an 18–19 year old base population extracted from the 2001 Census. The continuous variables were not converted to index scores. Clustering requires that data be measured on the same scale, and therefore all index scores created from the count data and the untransformed continuous variables were converted to z-scores. A z transformation is achieved using Equation (5.3).

Equation (5.3) –
$$z = \frac{x - \bar{x}}{\sigma}$$

This equation divides the difference between a case variable (x) from the mean of the total cases in the population by the standard deviation of all case values.

The choice of standardising with z-scores over other methods (such as range standardisation and log scores, used in the original OAC classification) is to prevent the scores being capped, and therefore allow possible outliers to influence the final classification as these may demonstrate important Higher Education behaviours. Because the units of clustering are larger than in OAC, i.e. Micro Groups rather than OAs, outliers created by aggregating data by small geographical areas are greatly reduced, and as such any outliers present in the data matrix should represent significant features which should be picked up in the final cluster assignments. The process of filtering small aggregation outliers from geographical features is helped in the clustering by weighting each case (Micro Group) by the total population for these areas, therefore reducing the weight/influence of those areas based on a smaller sample size.

A Higher Education specific geodemographic classification could feasibly serve a number of different purposes including marketing, extending access, widening participation or subject specific targeting. When selecting input variables these purposes were considered with the aim of creating a classification useful for a variety of tasks demanded by Higher Education decision makers. Variables selected are chosen to provide both indirect and direct indicators relating to those constructs which influence participation in Higher Education, and are reviewed in the following section. The classification measure created is likely to be suitable for profiling both access to Higher Education and disaggregated to predict course and institutional profiles. These applications are then tested in the next chapter.

The data available for this study comprise a subsection of the 2001 HESA database covering all students with English domicile studying within English institutions. This database contains of a variety of suitable variables for inclusion in the cluster analysis. The variables chosen are listed in Table 5.2 and the reasons for their inclusion follow.

Table 5.2   Higher Education input variables for the cluster analysis

Variable	Numerator	Denominator
Young participation rates	First year students aged 18–19	Census 2001 18–19
Average distance from student's home to institution	N/A	N/A
Average A-Level Score of students	N/A	N/A
Proportion of students from low social class groups	Undergraduate degree students from the three lowest social classes (IIIM, IV, V)	All undergraduate degree students
Proportion of degree course groupings*	Those studying undergraduate degree courses within groupings (A-X)*	All undergraduate degree
Proportion of different ethnic minority groupings**	Those undergraduate students from ethnic minority groupings	All undergraduate degree students
Proportion of students previously educated in Independent Schools in Years 12 and 13	Those undergraduate students who previously attended independent schools	All undergraduate degree students

* = Course Groups are defined in Table 5.3.

** = Ethnic minority groupings are found in Table 5.4.

Table 5.3   Course groupings

Course	Short Code
Medicine and Dentistry	A
Subjects Allied to Medicine	B
Biological Sciences	C
Veterinary Science, Agriculture and Related.	D
Physical Sciences	F
Mathematical and Computer Sciences	G
Engineering	H
Technologies	J
Architecture, Building and Planning	K
Social Studies	L
Law	M
Business and Administration Studies	N
Mass Communications and Documentation	P
Linguistics, Classics and Related	Q
European Languages, Literature and Related	R
Non-European Languages and Related	T
Historical and Philosophical Studies	V
Creative Arts and Design	W
Education	X

Table 5.4   Ethnicity groupings

Ethnicity Groups
White
Black or Black British – Caribbean
Black or Black British – African
Other Black background
Asian or Asian British – Indian
Asian or Asian British – Pakistani
Asian or Asian British – Bangladeshi
Chinese or Other Ethnic background – Chinese
Other Asian background

Young Participants

Young participants as defined by HEFCE are applicants accepted by a Higher Education institution who are aged 21 years or younger. Although this definition is most accurate in terms of national policy, the majority (~50%) of "young" applicants being accepted through UCAS during the period 2000–2004 were aged 18–19. When estimating participation rates, a base population of 18–19 year olds was extracted from the 2001 Census and compared with the average number of students of the same age in the HESA data shown to be attending Higher Education. If a 2001 Census base count of all residents aged 21 or less was used, this, of course would produce far lower "participation" rate figures, which are likely to be biased in a number of ways; for example, in overestimating participation rates in new estates with young families whose offspring are not yet old enough to participate in Higher Education.

Students and Distance

The use of distance travelled to accept a degree place is a proxy for geographic restrictions which are apparent in applicants from lower socio-economic groups, either through the financial cost of travel to a disparate institution or the social networks which bind them to their local community. Distance is measured in this analysis using a straight linear path between the accepting institution and the student's home; however, other distance metrics could also be used. Travel may, for example, be more prevalent in areas which demonstrate increased accessibility to transport networks. In this analysis, a straight linear distance was used as a pragmatic way of gauging relative variability between neighbourhoods, without more complex analysis involving transportation data. The coordinates for these locations are derived from the 2001 All Fields Postcode Directory. Once two pairs of xy coordinates were derived, where i is the home of the student who is attending j institution, Pythagoras Theorem was used to calculate the distance using Equation (5.4).

$$\text{Equation (5.4)} - \quad d = \frac{\sqrt{(x_i - x_j)^2 * (y_i - y_j)^2}}{1600}$$

Including distance in the classification aims to discriminate between those who may not participate because they are remotely located from institutions and those who may not participate because of their socio-economic profile.

Average A-Level Scores and Non A-Levels

In the 2001 HESA data, A-Levels were measured on a points scale, ranging from 10 points for a grade A to 2 points for an E. These scores were cumulative, so someone attaining AAE would score $10 + 10 + 2 = 22$ points. These scores were aggregated by each OAC Micro-Group to create median prior attainment scores. Prior attainment, particularly with regard to academic qualifications such as A-Levels have been seen as "key to the reaffirmation of middle class privilege in education and employment" (Leathwood and Hutchings, 2003:153) and so should provide a good discriminator of neighbourhood disadvantage. Where these scores are not recorded in the HESA data, the applicant will have qualified for Higher Education through a non A-Level qualification; these have been recorded separately in this analysis as a non-A-Level variable.

Social Class

In 2001 HESA data social class was measured on the Registrar General's Social Scale which groups occupation into seven different categories. These categories are:

- I Professional occupations
- II Managerial and technical occupations
- IIIN Skilled non-manual occupations
- IIIM Skilled manual occupations
- IV Partly-skilled occupations
- V Unskilled occupations
- VI Armed forces

Low rates of participation by the lower social classes have been documented ever since the Robbins Report (Robbins, 1963), and the extent to which these social barriers have been successfully addressed is debatable. In order for the classification to discriminate between the higher and lower social class groups, a variable was created from frequency of students within the groups IIIM, IV and V.

Degree Subject Chosen

The 2001 HESA data use the Standard Classification of Academic Subjects (SCAS) to aggregate individual courses into subject groupings. The extent to which different neighbourhood Types participate across these subjects is essential for both marketing and widening participation. The inclusion of the proportion of students within each subject grouping will improve how the classification discriminates participation rates between subjects.

Ethnic Groupings

The ethnic classification of students was included because membership of some ethnic minority groups has been observed to be associated with low participation and also the selection of certain subjects. Table 5.5 shows index scores created

Table 5.5   Index score cross tabulation between ethnicity and subject

Course	Asian	Black	Mixed	Other	White
A Medicine and Dentistry	246	63	124	188	88
B Subjects allied to Medicine	152	135	76	116	93
C Biological Sciences	68	76	100	83	107
D Veterinary Science, Agriculture and Related	26	16	36	27	117
F Physical Sciences	54	36	75	41	113
G Mathematical and Computer Science	190	137	93	137	85
H Engineering	123	137	96	130	94
J Technologies	72	64	73	100	106
K Architecture, Building and Planning	97	85	88	101	101
L Social Studies	92	158	110	99	97
M Law	178	147	108	140	88
N Business and Admin studies	165	159	94	126	87
P Mass Communication and Documentation	57	111	129	104	104
Q Linguistics, Classics and related	41	38	112	73	112
R European Languages, Literature and Related	24	30	121	76	116
T Non-European Languages and Related	42	31	195	116	111
V Historical and Philosophical Studies	28	20	90	70	116
W Creative Arts and Design	40	67	118	92	107

Source: UCAS Acceptances, 2005.

using the 2005 UCAS acceptance data to show course participation differences between ethnic groups. It can be seen, for example, that Asian students are almost two and a half times as likely to study Medicine and Dentistry then the student population as a whole.

Independent School Students

The inclusion of this variable is to increase performance of the classification to discriminate those areas which exhibit a strong propensity for students previously attending independent schools. One of the HEFCE performance indicators is based on the proportion of students coming from state schools, so the inclusion of this variable is in line with this performance measure. All variables included in the cluster analysis are detailed in Table 5.6 with their short codes which are used in the presentation of results.

Higher Education Case and Variable Preparation, Weighting and Correlations

Appropriate weighting of cases/data points is essential, and particularly so when clustering small geographic areas such as Output Areas, since a fluctuation of input scores created by low population counts can skew classification results towards outliers. By weighting the significance of cases in the cluster analysis by population size, those low population areas have less effect in the final cluster assignment and this contributes to the creation of more evenly sized clusters. For this reason, unweighted *k*-means is often used for outlier detection in multidimensional datasets, but in geodemographic applications, very low population counts in some clusters can reduce the efficiency with which it is possible to both describe and discriminate between societal groups. The educational classification will therefore be created with each Micro-Group having an assigned population weighting.

Before performing a cluster analysis, the data were explored to examine the correlations between the variables. It can be argued that highly correlated variables within a cluster analysis results in data redundancy and can have undesirable effects in final cluster assignment (Vickers, 2005). Harris et al. (2005) also discuss how it is important that included variables add new information rather than repeating what is already known. It is claimed that the methodology employed by Experian in the construction of Mosaic allows correlations to be maintained by weighting variables where necessary in order to improve the classification. Correlation between variables has the effect of adding extra weight to a particular dimension of the classification, and depending on which group of variables is being measured, could actually benefit the final assignment of clusters. For example, a key purpose of the Higher Education classification under development here is discrimination between areas that have high and low participation. One would expect the various factors which lead to these patterns of inequality to be highly correlated; for example, A-Level Points and Social Class. Both of these dimensions contribute towards low participation and therefore should be included. Their potential correlation reinforces an important dimension of the classification and as such should be allowed to manifest itself in the final cluster assignment. In a classification where a broad range

Table 5.6   Cluster analysis input variables

Variable
A-Level points
Distance travelled to attend institution
Lower Social Class
Black Caribbean
Black African
Other Black
Asian Indian
Asian Pakistani
Asian Bangladeshi
Chinese
Other
Medicine and Dentistry
Subjects allied to Medicine
Biological Sciences
Agriculture and related subjects
Physical Sciences
Mathematical Sciences and Informatics
Engineering
Technology
Architecture, Building and Planning
Social Studies
Politics and Law
Business and Administrative Studies
Mass Communications and Documentation
Linguistics, Classics and related subjects
European Languages, Literature and related subjects
Eastern, Asiatic, African, American and Australasian Languages, etc.
Humanities
Creative Arts
Education
Combined and general courses not otherwise classified
No A-Level Points (i.e. non A-Level Qualifications)
18–19 Young Participation Rates
Independent School

of variables are included to inform a number of roles, the correlation between small subsections of these data reduces the need for variable weighting. The converse argument is that these dimensions could be artificially created through the use of effective weighting. The main issue with variable weighting to inform a classification is that these choices are inherently influenced by those a priori models held by the researcher of how the final classification will represent the data. If a classification builder wished a particular dimension to be over represented, for example the existence of wealthy Asians versus poorer Asians, this could quite easily be achieved through variable weighting. The Output Area Classification (OAC) minimised the inclusion of correlated variables and did not use weights. The bespoke educational classification developed here only includes relevant variables, some of which are correlated, and like OAC did not use variable weighting for the reasons discussed above.

In order to understand the correlation within the input dataset, a matrix of all input variables was created using a population weighted Pearson Correlation Coefficient. Highly positively correlated variables included A-Level Points, independent schools, distance travelled to accept a place and young participation. As one would expect, each of these variables is highly negatively correlated to low Social Class and no A-Levels. These patterns are unsurprising and refer to a core component of the classification, to discriminate between those areas of high and low participation. One would also expect these variables to correlate with some subjects as entry grades vary between subject groups, subjects appeal to different people and subjects are not evenly distributed across Higher Education institutions. For example there is a high negative correlation between low Social Class and participation in JACS Group A Medicine and Dentistry.

How Many Clusters Should an Educational Classification Have?

The k-means algorithm clusters the input data matrix into the k number of clusters are specified by the researcher. Therefore, unless a prior model of how many groups should exist within the dataset is known, a method of selecting a sensible cluster frequency is required. As mentioned earlier in this chapter, one method of doing this has been demonstrated by Debenham (2001), and entails running the k-means algorithm for multiple iterations of k and plotting the average distance between the data points and their closest cluster centroid. These charts show the homogeneity of each cluster solution against the number of clusters. The higher the number of clusters, the smaller the mean distances between the data points and the nearest cluster centroid. The charts thus illustrate the trade-off between mean distance and classification complexity. Debenham (2001) conducts this analysis by running only a single cluster analysis for each k value. This has the disadvantage described earlier that the k-means algorithm is sensitive to the location of initial seeds; a problem that can be largely circumvented through repeated analysis using multiple initial seed values. Debenham (2001) selects a final k value based on interpretation of apparent breakpoints in the plot of cluster homogeneity against the number of clusters. However, without re-running the cluster analysis, these observations may be anomalies based on inappropriately selected initial random seeds. Although this method is useful in principle, it needs to be adapted in order to provide more robust results.

The method adopted in this study builds on Debenham (2001) and runs the algorithm for k_{n-2} models where n is the total number of Micro Groups within the dataset (176). However, in order to improve the confidence with which the trade off between cluster homogeneity and numbers of clusters is made, each iteration of k was re-run 10,000 times. The median, minimum and maximum distances and overall R-Squared were averaged over the 10,000 iterations for each k value and are graphed in Figure 5.7 and Figure 5.8. In these Figures, the dark line represents the median and the grey whiskers the minimum and maximum values. The systematic location of the median towards the maximum values in Figure 5.7 indicates that there are more iterations of the model close to the maximum performance than the minimum performance.

These graphs show that the R-Squared increases with the number of clusters specified, although not in a linear fashion. Furthermore, as k decreases, so the variability of the R-Squared increases, providing further justification of the need for multiple model runs to attain robust information, particularly at lower values of k. The increased variability in R-Squared at most of the lower k values is caused by the grouping of the data points into smaller aggregations, and this corresponds to a greater probability that a final case allocation can switch between clusters (since this increases the variability of the final classification performance). Furthermore,

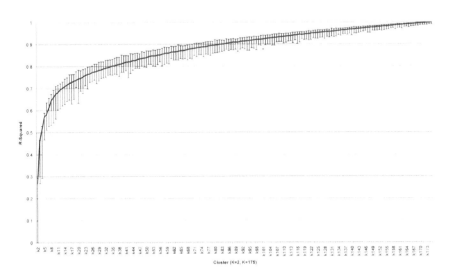

Figure 5.7   Cluster performance measured by R-Squared scores
(k = 2–175)

these results show that further increases in *k* result in successively smaller improvements in *R*-Squared, and at the crudest aggregations, much information is lost. The *R*-Squared plots are very useful for selecting an appropriate cluster number for the dataset, as the loss in performance of the classification can be assessed and compared for each reduction in *k*. The difference in *R*-Squared scores created by increasing the frequency of *k* from *n* to *n+1* is shown in Figure 5.9.

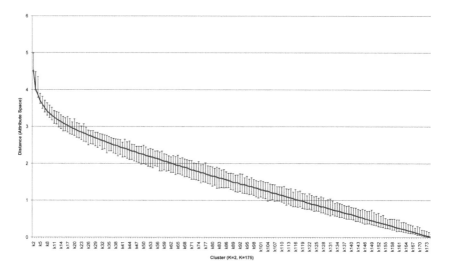

Figure 5.8   Cluster performance measured by distance scores (*k* = 2–175)

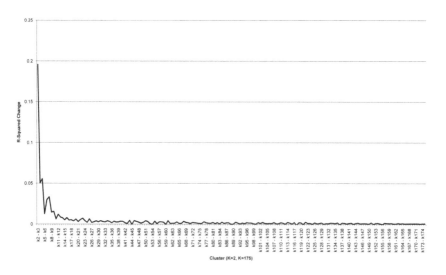

Figure 5.9   *R*-Squared difference scores

The improvement in *R*-Squared is reasonably even and low beyond $k = 40$ and improves considerably, although sometimes erratically, for all values of $k \leq 17$.

Geodemographic classifications are created hierarchically and consist of a series of aggregations. This allows end users to have greater flexibility over the detail they can present and also the number of groups into which their own data are divided. Having a classification with a large aggregate level is useful when profiling data from a small population, e.g. an unpopular course against all courses at a university. The exact number of clusters and levels varies between classification providers and these are summarised in Table 5.7. With the exception of OAC, little justification is given to why particular levels of detail are chosen.

Table 5.7　　Classification levels

Classification System	Clusters in Level 1 (<12 Clusters)	Clusters in Level 2 (>=12, <50 Clusters)	Clusters in Level 3 (>50 Clusters)
Mosaic 2001	11	–	61
Cameo	10	–	58
ACORN	5	17	56
PRiZM	–	16	60
Super Profiles	10	40	160
OAC	7	21	52

Source: Adapted from Vickers, 2005.

The classification used most prevalently in Higher Education since 2001 has been Mosaic Types with two hierarchies of 11 and 61 clusters. It was preferable to keep the new bespoke education classification in line with similar levels and cluster frequencies in order not to confuse potential end users with radically different aggregations.

Creating the Bespoke Educational Classification

The final classification should be fit for a range of purposes demanded by Higher Education decision makers. Most importantly, the classification should provide an effective way of discriminating between those areas of high and low participation in aggregate, and also disaggregated by course types to allow more specific targeting strategies. As in commercial geodemographic classification, it is useful for a bespoke educational classification to have multiple levels which create flexibility when analysing target groups of different sizes. The classification will also most prevalently be used to discriminate between 18–19 year olds as they form the majority of the Higher Education demographic, and therefore the classification should aim to have a relatively even distribution of this age range between the final cluster assignments.

The selection of variables was detailed earlier, but it is also appropriate to investigate the most appropriate value of *k*. In line with the observation from Figures 5.7 and 5.8, that there are no discrete steps in performance and thus no single appropriate value of *k* in statistical terms, as such, 10,000 separate cluster analysis were run from *k*=50 to *k*=65. These values of *k* are in a similar range to the finest level of aggregation shown previously in Table 5.7. The median, minimum and maximum R-Squared results are presented in Figure 5.10.

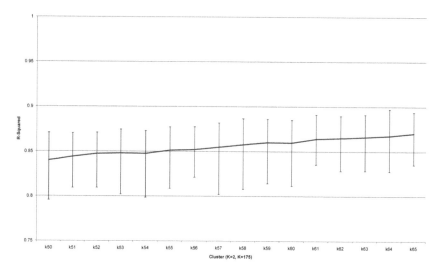

Figure 5.10   *R*-Squared results from 10,000 cluster runs

Each of these assignments of *k* appears to perform well at discriminating within the input data matrix and, as demonstrated in the earlier exploratory analysis, the minimum and maximum bars further illustrate the need to optimise each *k* allocation. The total and 18–19 year old population from the 2001 Census were then aggregated into the *k*=50 to *k*=65 models in order to ensure that no outliers of this key target population had been created in the clustering process. The model demonstrating the most even distribution of 18–19 year olds across the new clusters was *k*=53 (see Figure 5.11) and therefore was chosen as the final model. It should, however, be noted that the distribution of 18–19 year olds is still skewed. The dark line is drawn at 1.89 which would divide the principal applicant group equally between the 53 clusters (i.e. 100/53). The uneven distribution of household and population counts is characteristic of most geodemographic classifications and in the Mosaic commercial classification the percentage assignment of total households to Mosaic Types ranges from 0.17% – 3.82%.

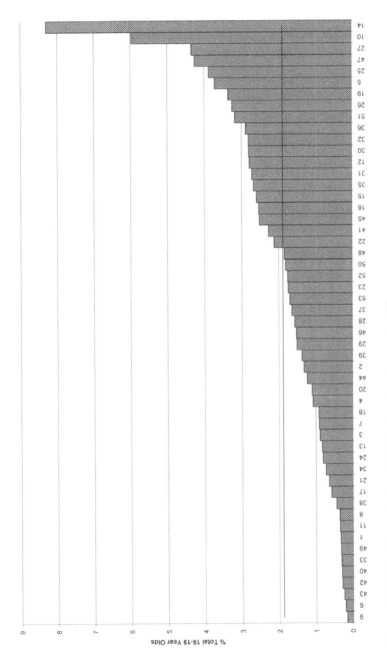

Figure 5.11 18–19 population distribution by cluster ($k = 53$)

The Educational OAC Type level classification was therefore defined as comprising of 53 clusters. However, as discussed earlier, it is often useful to have a second hierarchy in classification which aggregates the finest level into larger groups. A second type of clustering algorithm is used to aggregate the 53 clusters (Types) into the larger aggregation of Groups. The Ward (1963) method assesses the loss of variance that would be associated by merging clusters together when those which are amalgamated minimise the "increase in information loss" (Everitt, 1993:65). Information loss is defined by an error sum of squares criterion (ESS).

Equation (5.5) – $$ESS = \sum_{i=1}^{n} \left(x_i - \overline{x} \right)^2$$

Equation (5.5) measures the total sum of the squared deviation for all variables from each of the 53 Types to the means of the clusters to which they might be assigned. At each step in this process, the algorithm iterates through all possible unions for the 53 Types, and at each pairing, an assessment is made using this equation to calculate the increase in the error sum of squares. The union with the lowest increase in error sum of squares is actioned, and the process continues for further iterations until all 53 Types have been progressively combined into a single cluster. The hierarchical organisation of Types into Groups can have multiple arrangements. The performance of these Group classifications for predicting a target variable (e.g. participation) will depend upon the level of correlation between the variables used in cluster analysis and the target variable. Harris et al. (2005) suggest that group level classifications should ideally have populations no lower than 4% and no greater than 20%, and also contain between two and seven constituent clusters (see Figure 5.12).

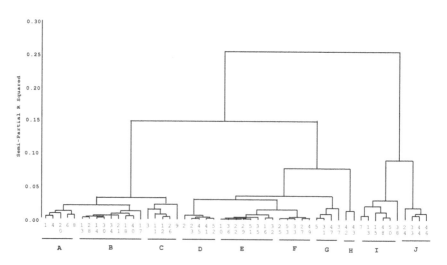

Figure 5.12   Clustering dendrogram

The y axis records the proportion of variance accounted for by joining successive clusters together, and thus, when choosing the Group level a balance is sought between the aggregation which provides a degree of separation between grouped clusters, and one which provides a level of division that is analytically useful for end users. A 10 Group level classification was selected. The squashed bases indicate that the new classification is parsimonious at Group level, and thus suggests that the Group level classification may be useful for assessing participation rates.

Cluster Descriptions

In order to visualise how the classification discriminates between OAs, a map has been drawn of a part of North London which represents an area ranging from very affluent to very deprived neighbourhoods within a relatively small geographical area (see Figure 5.13).

Figure 5.13   Educational OAC Groups in North London

To examine the Higher Education characteristics of the Group level classification, UCAS acceptance data for 2000–2004 were appended to the output area lookup table and index scores (base score 100) were calculated from these aggregations across a range of variables. The base used in the calculation was the proportion of all acceptances over the period 2000–2004. Using these data, a range of descriptions have been written to describe the clusters. The variables for which these index scores were calculated included:

- Propensity for course level participation
- Propensity to attend a Russell Group[3] institution
- Propensity for OA Group to appear in specific POLAR participation bands
- Propensity for the OA Group to contain particular Mosaic Groups

These statistics provide useful material for end users to describe the general characteristics of students within each of the educational OAC neighbourhood Groups, and also a method of external validation through cross classification with Mosaic Groups.

Group A

Group A are the very lowest participation neighbourhoods, generally found within urban areas outside London. Students who do attend university from these areas have a high propensity to study either Education or a Biological Science. Attendance at Russell Group institutions is half the national average.

Group B

Group B neighbourhoods are low participation, again found predominantly outside London. Those students who do attend university also have a high propensity to study Subjects Allied to Medicine, Education or Creative Arts and Design. Few students in these areas attend Russell Group institutions.

Group C

Participation in Group C neighbourhoods is low, but markedly higher than Groups A and B. These neighbourhoods are predominantly found on the outskirts of larger urban areas in the South and Midlands of England. A high proportion of participating students study Mathematics and Computer Science. Participation in Russell Group institutions is low.

---

3   The Russell Group is an association of leading UK research-intensive Universities whose membership include: University of Birmingham, University of Bristol, University of Cambridge, Cardiff University, University of Edinburgh, University of Glasgow, Imperial College London, King's College London, University of Leeds, University of Liverpool, London School of Economics and Political Science, University of Manchester, Newcastle University, University of Nottingham, Queen's University Belfast, University of Oxford, University of Sheffield, University of Southampton, University College London, University of Warwick.

Group D

Group D neighbourhoods are predominantly rural, with a moderately high participation rate. As one might expect of rural residents, many students study Veterinary Science or Agriculture. Other popular course choices include Engineering, Technologies, Architecture/Building/Planning and both European and Non European Languages. Students from this Group are 25% more likely to attend a Russell Group institution than the national average.

Group E

Group E neighbourhoods are found across all main urban areas (although only the periphery of London). They have average to low participation rates, but Education is a popular subject of study. Other subjects that are reasonably popular include Biology, Physical Sciences, Mass Communications/Documentation and Creative Arts/Design. This Group has an average propensity to attend a Russell Group Institution.

Group F

Group F are frequent participants in Higher Education, and are found in both rural and urban areas across England, predominantly in affluent neighbourhoods. They exhibit no strong course preferences, but are 17% more likely than the national average to attend a Russell Group institution.

Group G

Group G populate affluent neighbourhoods in the suburbs of many large urban areas, particularly in London and the South East of England. Students from these areas are very likely to go to university, and have a tendency to study Medicine, Architecture, Building and Planning, Social Studies, Languages, Philosophy or History. Students from these neighbourhoods are 50% more likely to study at a Russell Group institution than the national average.

Group H

The predominantly rural neighbourhoods in Group H are the most likely of all Groups to participate in Higher Education, and students from these neighbourhoods are 66% more likely to study at a Russell Group institution than the national average. Subjects which are popular with students from these areas include Veterinary Science, Physical Sciences, Engineering, Social Studies, Business Studies, Languages, History and Philosophy.

Group I

Group I neighbourhoods have average to low participation rates, and mainly appear in urban areas, especially Inner London. Students from these areas typically study Computer Science and Mathematics and a range of other subjects. These students are unlikely to attend a Russell Group Institution.

Group J

Group J neighbourhoods have average rates of participation. They are mainly found in large urban areas across England, especially in the suburbs of London. Many students study Medicine or Subjects Allied to Medicine, Mathematics, Computer Science, Law, and Business. Students from these areas are 30% less likely than the national average to attend a Russell Group Institution.

Conclusion

This chapter has demonstrated a pilot method by which bespoke classifications for a particular domain or application can be created, using public sector data sources. The motivation for this analysis lies in the observation that typologies created by commercial classification providers supply no evidence to justify why the inclusion of data relating to private consumption of goods is appropriate for predicting public consumption. Furthermore, the exact nature of the weighting schemes and data used to derive such commercial classification systems is closed to the public, which is of concern in public sector applications which may apportion real life chances, rather than just material consumer offerings. The addition of Higher Education sector data is seen as a positive step beyond use of generic and re-labelled classification for purposes for which they were not originally designed, and a challenge to the implied assumption that individual use of public services, such as education, should directly correspond to the ways in which consumers use private goods and services. This work also responds to concerns that the data inputs used to create generic commercial geodemographic classifications come from disparate private sector and closed sources; their provenance is often unknown; and that the assumptions used to create such classifications cannot be scrutinised or tested by end users. The negative potential social implications of using such classifications in areas of public service provision should not be under-estimated, and potentially significantly reduce the life chances of stakeholders in public services. The methodology has shown how a classification built using the 2001 Census can be refined for a specific purpose through the augmentation of sector specific data. The final classification consists of 10 Groups and 53 Types.

Chapter 6

Evaluating Geodemographic Performance for Profiling of Access

Evaluation of Discrete Classification

It was argued in the previous chapter that when creating geodemographic classifications, the assignment of areas into clusters can have numerous outcomes through altering the numbers of divisions, the number of hierarchies in the typology, the geographic scale used to create the clustering units, and finally the multiple possible arrangements of spatial clustering units within an "optimised" cluster model. The lattermost of these issues relates to how the k-means algorithm can have multiple locally optimised models, depending on the location of the initial seeds used as input into the cluster analysis. As discussed previously, there are numerous classifications used in applications both across the public and private sectors, and the use of these classifications has thus far remained uncritical. However, this chapter aims to evaluate the performance of a number of discrete models for Higher Education data segmentation, using a series of evaluation methods.

The evaluation of discrete classifications can be considered both quantitatively and qualitatively. A qualitative evaluation focuses on those attributes which are considered to make one classification more fit for purpose than another. Indeed, Leventhal (1995:6) argues that "while numerical measures provide helpful summaries, they cannot evaluate the usefulness of a discriminator as opposed to the statistical significance". Qualitative evaluations could be conducted for a particular application (e.g. creating targeting strategies to increase a course's recruitment), or as an overall assessment for use of a classification within a particular sector (e.g. HE). These types of evaluation are subjective, and as such should not be used in isolation to determine the suitability of a classification. They nevertheless do provide a good framework to learn about the numerous discrete classifications which may be suitable, and also provide a method of choosing which classifications could be used in more rigorous empirical testing.

The data used for the quantitative evaluation are derived from 2004 UCAS acceptances for England to enable the broadest range of classification to be compared. Using a Total Weighted Deviation (TWD) method, the predictive performance of classifications may be examined for courses of Higher Education and also for aggregate institutional profiles. These represent two core analyses that an institution might complete in order to market or extend participation to people living within specific neighbourhoods. A second evaluation method compares the performance of the classification to identify and therefore target neighbourhoods which are unlikely to supply the traditionally aged (18–19) participants to study degrees in Higher Education. The classifications compared in this evaluation are those which were made available to the researcher, which included the National Statistics Output Area Classification, Experian's Mosaic, CACI's ACORN, the bespoke classification created in the previous chapter and the Index of Multiple Deprivation (IMD). The IMD is measured on a continuous scale, but for the purpose of this analysis the classification was divided into 10 deciles. For the TWD analysis only, the National Statistics Socio-Economic Classification (NS-SEC) and the HEFCE Participation Groups (POLAR) are also evaluated. They are excluded from the aggregate participation performance analysis as 18–19 year old base scores could not be derived for a comparable time frame.

Qualitative Analysis: What Makes a "Good" Geodemographic Classification?

The first stage in a qualitative evaluation is to identify a set of criteria through which the classifications will be assessed. These might include:

- Frequency of clusters
  - More clusters may be considered advantageous as this would provide a finer level to neighbourhood profiles.
- Hierarchy of clusters
  - An increased number of hierarchies demonstrates that the classification has flexibility when analysing target groups of varying population sizes. Partitioning a small population by a typology consisting of many categories would create unreliable/unstable index scores.
- Geographic scale of classification
  - A classification which identifies neighbourhoods (clustering units) at a fine scale such as the individual, household or unit postcode is, all other things being equal, of greater usefulness than one available only for coarser aggregations.
- Suitable labelling of clusters
  - The clusters have memorable names which communicate their key attributes to a target audience.
- Suitable "Pen Portraits" to describe the clusters
  - Full contextual descriptions on the attributes which describe people living within the neighbourhoods defined by the clusters.
- Appropriate multimedia and imagery
  - Classification might be supported by a suite of multimedia such as video, audio, images and graphs. These enable quick reference to the key attributes of the classification.

- User groups
    - The classification is supported by a user group, perhaps with members active within specific domains of use, e.g. crime.
- Pedigree within the sector
    - The classification has been used within the sector, providing a framework for comparative analysis.

These criteria can be used qualitatively to assess the application of geodemographic classifications in Higher Education. A summary matrix of the application of these assessment criteria to selected geodemographic systems is presented in Table 6.1.

Although this table suggests criteria through which classifications could be ranked (e.g. the summation of the "Y" attributes down a column), it would be limiting to exclude the use of a specific classification on this basis, not least because the selection of both evaluation criteria and completion of the matrix are subjective. Although providing a broad brush evaluation, this method of evaluation needs to be substantiated through more comprehensive empirical analysis. A qualitative matrix is better interpreted as a method of deriving the strengths and weaknesses for a particular segmentation solution prior to more detailed quantitative assessment of discriminatory power.

Methods of Quantitative Evaluation

Quantitative evaluation will be complete using a total weighted deviation (TWD) method to examine the predictive power of classifications across a range of variables; and Lorenz Curves (Gains Curve) and Gini Coefficients to examine a single application of predicting the participation rates of young entrants to Higher Education. In order to create a measure of young participation as a function of the underlying base population, the estimated frequency of 18–19 year olds from the census was required at postcode level. Data from the 2001 Census are distributed at its finest level within Output Areas (OA), which although appropriate for OA level classifications such as OAC, others such as Mosaic and ACORN classify neighbourhoods at the level of unit postcode. An additional output from the 2001 Census available from the Office of National Statistics (ONS) is a postcode file which disaggregates the total, male and female populations by unit postcodes. Using the distribution of the unit postcode level total population frequency (R) within an Output Area (OA) the 18–19 year old data (T) could be apportioned, thus estimating frequency of 18–19 year olds at postcode level (Y) based on an assumption that this demographic is distributed across each OA uniformly (see Equation 6.1). Because these data would later be aggregated into the classification categories, the data were not rounded, and as such a unit postcode could apparently contain non-integer values with respect to 18–19 year old persons.

Table 6.1   Qualitative evaluation metric for geodemographic classification

	Mosaic Type	Mosaic Groups	OAC Sub Groups	OAC Groups	OAC Super Groups	Educational OAC Types	Educational OAC Groups	ACORN Category	ACORN Group	ACORN Type
Frequency of clusters	61	11	52	21	7	53	10	5	17	56
Hierarchy of clusters	2			3		2			3	
"Fine" geographic scale	Y	Y	N	N	N	N	N	Y	Y	Y
Clusters are named	Y	Y	N	Y	Y	N	N	Y	Y	Y
"Pen Portraits"	Y	Y	N	Y	Y	N	N	Y	Y	Y
Multimedia	Y	Y	Y	Y	Y	Y	Y	Y	Y	Y
User Group	Y	Y	Y	Y	Y	Y	Y	N	N	N
Pedigree	Some	Some	N	N	N	N	N	Some	Some	Some

Equation (6.1) –
$$y_i = T_{OA} \times \frac{R_i}{\sum\limits_{i=1}^{n} R_i}$$

Apportioning population in this way is computationally intensive because of the size of the files being manipulated. The apportioning algorithm was run iteratively for every OA using the statistical software SAS, taking a number of days to run through to completion. This type of analysis requires a significant level of statistical and programming ability and is not an analysis that many end users of geodemographic classifications would be either able, or willing to complete. For specific applications dealing with a sub population group this makes OA level classifications far easier to use than those based on postcodes. To circumnavigate these issues, commercial vendors will often supply a software application where these analyses are predefined.

Social Similarity, Clustering Scales and Indices of Dissimilarity

The previous chapter presented an argument that attribute space can be partitioned in multiple arrangements and therefore that geodemographic classifications do not present a definitive or uncontested representation of reality (Gordon, 1980). However, despite these multiple possible cluster solutions, it appears that geodemographic classifications do share common characteristics of social similarities within the groups of classification typologies; that is, clusters have shared characteristics within their common attribute space. Table 6.2 shows index scores and percentages which demonstrate the degree of similarity between Mosaic Groups and OAC Super Groups. While a number of Mosaic Groups have significant overrepresentation within particular OAC Super Groups, others show less extreme overrepresentation with a more even distribution. A commonality between those which are more evenly distributed is that they are predominantly coded by Mosaic as people living in more affluent areas. If reality is assumed to be accurately represented by the Mosaic Classification, it could be interpreted that these more affluent neighbourhoods are less well defined in the OAC classification, i.e. by the census. However, what is encouraging for those using OAC in applications targeting less deprived areas (as extracted from the Mosaic descriptive material), is that this classification shows a great deal of similarity to Mosaic across these neighbourhoods.

A further difference between these two classifications is the scale of geographic units used as input to the cluster analysis. Mosaic is defined at the scale of the unit postcode and OAC at 2001 Census Output Area. Harris et al. (2004:219) recognises that "there is no optimal scale for classifying neighbourhoods. Consumer behaviour within some product categories is better predicted using demographic data for areas more geographically extensive than Census output areas, while for others

Table 6.2 Index scores and percentages (within brackets), comparing Mosaic Groups with OAC Super Groups

Mosaic\OAC	Blue Collar Communities	City Living	Countryside	Prospering Suburbs	Constrained by Circumstances	Typical Traits	Multicultural
Symbols of Success	1 (0.2)	127 (8.7)	108 (23.4)	277 (57)	1 (0.1)	48 (8.9)	20 (1.7)
Happy Families	62 (8.2)	27 (1.8)	43 (9.3)	203 (41.7)	16 (1.7)	186 (34.5)	31 (2.6)
Suburban Comfort	23 (3)	7 (0.5)	93 (20.1)	225 (46.4)	4 (0.4)	126 (23.4)	72 (6.2)
Ties of Community	111 (14.6)	78 (5.3)	19 (4.1)	39 (7.9)	56 (5.9)	238 (44.2)	209 (18)
Urban Intelligence	0 (0.1)	747 (51.1)	1 (0.3)	3 (0.5)	8 (0.8)	101 (18.8)	330 (28.4)
Welfare Borderline	16 (2.1)	196 (13.4)	0 (0)	0 (0)	469 (49.7)	0 (0)	404 (34.7)
Municipal Dependency	381 (50.2)	0 (0)	0 (0)	0 (0)	397 (42.1)	0 (0)	89 (7.7)
Blue Collar Enterprise	511 (67.2)	1 (0)	8 (1.7)	2 (0.4)	186 (19.7)	26 (4.9)	71 (6.1)
Twilight Subsistence	91 (12)	46 (3.1)	7 (1.5)	2 (0.4)	699 (74.2)	21 (3.9)	57 (4.9)
Grey Perspectives	28 (3.7)	180 (12.3)	117 (25.3)	104 (21.5)	49 (5.2)	166 (30.7)	14 (1.2)
Rural Isolation	1 (0.1)	0 (0)	452 (98)	7 (1.4)	0 (0)	2 (0.4)	0 (0)

the appropriate granularity is as low as unit postcodes". Therefore, Mosaic may demonstrate a finer level of detail about neighbourhoods beyond the aggregate characteristics of OA through the imputation of higher resolution data. Again under the assumption that Mosaic depicts an accurate representation of reality, it is possible to consider the effects of using a larger areal aggregation to assign a typology. This is a useful tool to assess the level of information loss that OAC may incur because it is created only from data aggregated at larger geographical units.

An algorithm was created which counts the frequency of different Mosaic Groups at unit postcode level within each UK OA. Those postcodes for which Mosaic has no information (e.g. Business addresses) were excluded from this analysis, creating a dataset of around 1.4 million postcodes. For each Output Area the frequency of different Mosaic Groups was appended to the total 2001 Census population within these areas. When aggregated by the frequency of Mosaic Groups, as shown in Table 6.3, this provides a guide to the magnitude of people who may be misclassified by using a classification which appends to a larger geographical unit.

Table 6.3   Total population within OA by frequency of different Mosaic Groups

Different Mosaic Group Frequency	Total Population	Percentage Population
0*	785,009	1.3
1	28,525,957	48.6
2	21,838,386	37.2
3	6,553,279	11.2
4	912,243	1.6
5	68,253	0.1
6	2,413	0.0
7	408	0.0

Note: * The population within OA that have 0 postcodes are OA in which all postcodes are classified as unknown in the Mosaic classification.

Around 48.6% of the total population live within OAs that are considered socially homogeneous at the unit postcode level by Mosaic Groups. A further 37.2% live within OAs which contains two different Mosaic Groups. Only a very small proportion of the total population live within OA which are very diverse at the level of the unit postcode.

The postcodes within these OAs can be profiled to assess the types of neighbourhoods which are more likely to cluster at OA level (see Figure 6.1). It could be that the clusters defining these areas are identified from within the attribute space predominantly by census variables which are disseminated at OA, or, that these areas are defined by neighbourhoods which typically may show greater homogeneity. The latter interpretation is supported by a large overrepresentation of the Type "Just Moving In" from within the Group "Happy Families". These areas are typically newly built houses within homogeneous estates where one may expect a greater degree of social similarity across an area typically measured by an OA.

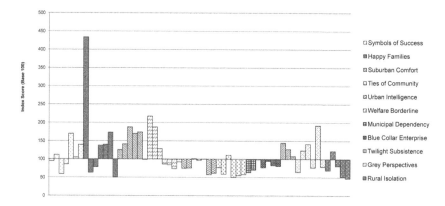

Figure 6.1   Homogeneous Output Areas by Mosaic Types and Groups

Heterogeneous OA with two Mosaic Groups have been shown to account for a significant proportion of the total population (37.2%). The following analysis considers the magnitude of the distribution between the two groups. If an OA contained a large proportion of postcodes from one Mosaic Group and a single postcode of another group, this could perhaps be considered less significant than a OA in which there was an even split in unit postcodes between two neighbourhood groups. Alternatively, it could be argued that these micro level differences between unit postcodes, no matter how small in magnitude, are a key discriminator that makes classification disseminated at the unit postcode level more powerful.

A subset of the data was created which consisted of those OA containing unit postcodes split across two Mosaic Groups. The proportion of the largest of these groups was calculated as a percentage of the total. However, these scores would be influenced by the size of the total frequency of postcodes within an OA. For example, an OA which contained only two postcodes could only ever be assigned a single score of 50%. The range of scores that can be assigned to an OA will therefore increase as the frequency of postcodes increases (See Table 6.4).

Table 6.4   Possible homogeneity scores

Postcodes frequency within OA	Frequency of maximum possible scores	Those scores possible
2	1	50%
3	1	60%
4	3	50%, *75%*
5	2	80%, 60%
6	2	83.33%, 66.66%, 50%
7	3	85.71%, 71.43%, 57.14%
8	4	87.50%, 75.00%, 62.50%, 50.00%
9	4	88.89%, 77.78%, 66.67%, 55.56%
10	5	90.00%, 80.00%, 70.00%, 60.00%, 50%

The frequency of OAs and the number of different postcodes within them is shown in Figure 6.2.

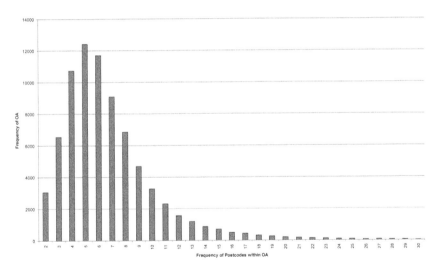

Figure 6.2   Frequency of postcodes within Output Areas

One could therefore expect certain percentage scores to be more overrepresented based on their probability of appearing within the dataset, e.g. 50% could be possible for every even number of postcodes. Despite this caveat the analysis still provides a useful measure of OA level neighbourhood heterogeneity. The distribution of the maximum percentage scores created by this analysis is shown ranked by order of frequency in Figure 6.3. With the exception of 50%, the top five OA frequencies are all over three quarters allocated to a single Mosaic Group. In the top five percentage groups ranked by frequency of OA, there are scores of 80% and 83% of postcodes allocated to a single Mosaic Group. These findings are encouraging for users of OA level classifications, as it appears that the loss of postcode level information is minimal within the majority of OAs split into two or fewer Mosaic Groups.

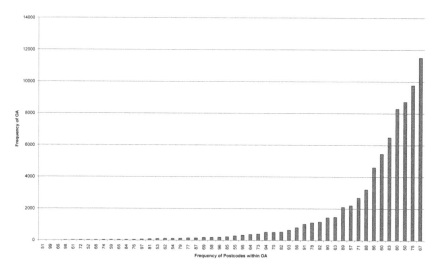

Figure 6.3   Distribution of maximum percentage scores

The analysis was run again on those OAs which contained unit postcodes that were split into three Mosaic Groups (approximately 11.2% population) (see Figure 6.4). It can be seen that the ranges of Mosaic Groups within these OAs are more heterogeneous than those OA with dichotomous Mosaic Groups; however there are still a large number of OA with 80% and 70% of their postcodes assigned to a single Mosaic Group.

   These analyses have started to demonstrate at what scale it is appropriate to create and disseminate geodemographic classification. There appears to be evidence suggesting relatively little heterogeneity between postcode level Mosaic Groups within and OA, and this supports the argument that the OA may be an appropriate scale at which to categorise neighbourhoods. Voas and Williamson (2001) argue that general purpose classification built at similar small geographies (Enumeration Districts) conceal hidden diversity at a finer scale, however, this can be reduced through task specific classification, as has been argued in the previous chapter through the creation of a bespoke educational classification. The alternative

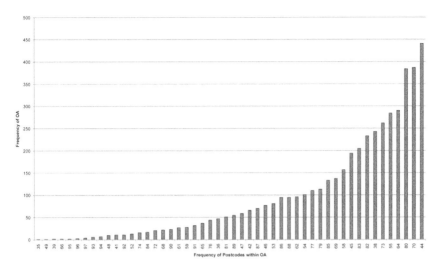

Figure 6.4   Distribution of maximum percentage scores

argument, and one which is not supported by the findings of this analysis, is the proposal made by Harris (2001) that finer scale information, such as those available in lifestyle databases at individual or household level, may produce more satisfactory classification to discriminate between areas. However, as Vickers (2006) discusses, these finer scale geographies are more prone to error induced through ageing of the data. The finer the scale at which input data area collected, the more temporal refreshing is required to maintain them in ways which are accurate and safe to use. At a larger scale, such as OA, small changes in the population may not alter how an area should be most appropriately classified, as the average characteristics of those resident still within the area will predominantly remain aligned to the classification representing these aggregate characteristics. Furthermore, those micro-level data used by commercial classification builders often originate from sample surveys which are not representative of the total population, and therefore the implied precision of a classification at unit postcode level accuracy may be superfluous. Indeed, the use of these data may explain why a minority of OA in this analysis were found to be very heterogeneous, with one OA in Northern Ireland containing 7 different Mosaic Groups, perhaps caused by erroneous imputation of sample data or the unavailability of household level databases. Finally, when Mosaic Groups were compared with OAC Super Groups, there was a high degree of social similarity between the classifications indicating comparability between how the two classifications categorise neighbourhoods. These effects seem to occur despite the modifiable areal unit problem (MAUP) of scaling (Openshaw, 1984; Wrigley et al., 1996) between the postcode and Output Area. However, these similarities were less pronounced in the more affluent Mosaic Groups, and may indicate a possible weakness in OAC as these neighbourhoods would be more difficult to identify.

Indices of Dissimilarity

A further method by which the heterogeneity within OAs can be gauged is through the use of diversity indices. These are established techniques in ecology to measure the heterogeneity of species within a defined area. These techniques can be adapted to provide a further method of measuring the homogeneity of Mosaic Groups within OAs. The Simpson Diversity Index (Simpson, 1949) measures the balance of neighbourhood groups within an OA and the scores range between 1 and 0. A more heterogeneous OA would contain a number of different Mosaic Groups and have a score closer to 0, whereas a homogeneous OA would have fewer Groups and a score closer to 1. The Simpson Diversity Index (D) in OA e is calculated by summing the square root of m Mosaic Groups ranging from $i=1$ to $=11$ which correspond to each of the 11 Mosaic Groups (See Equation 6.2).

Equation (6.2) – $$D_e = 1 - \sum_{i=1}^{11} \left(m_{ei}^2 \right)$$

Once scores have been created for each OA, they may be mapped to examine the degree of OA level homogeneity (see Figure 6.5). It can be seen that in urban areas there is a tendency for OAs to have a higher diversity (low score) with rural areas such as North Devon, Mid to North Wales, North Lancashire and the majority of Scotland all having high scores, i.e. low diversity. One exception to this broad pattern is that Northern Ireland appears quite diverse, even in areas which would be considered predominantly rural. Using the OA diversity scores, it is possible to aggregate these into the OAC typology in order to assess whether particular clusters demonstrate heterogeneity within their assigned OA. Because OAC clusters vary in size, these aggregated scores were transformed into index scores using a base of the total OAC distributions. Thus, Figure 6.6 represents those neighbourhoods by the OAC classification which typically show a greater heterogeneity at unit postcode level within the OA. A high index score indicates an area which is heterogeneous, and a low index score more homogeneous. There are differences both between and within the Super Groups and the following section presents a number of interesting findings. The Super Group "Constrained by Circumstances" contains three Groups. The first ("Senior Communities") and last ("Public Housing") of these Groups are both more homogeneous while the middle Group ("Older Workers") are more scattered alongside other Groups within OAs. From within the Super Group "Blue Collar Workers" the last Group "Older Blue Collar" are predominantly more heterogeneous, indicating a demographic dimension which relates to this diversity at postcode level within these areas. Within the Super Group "Countryside", the Group "Agricultural" is shown to be more homogeneous. The other two groups within this Super Group are areas which are more urban, such as small villages or towns. It is hinted in Harris et al. (2005) that Experian may re-cluster non urban areas in a separate analysis in

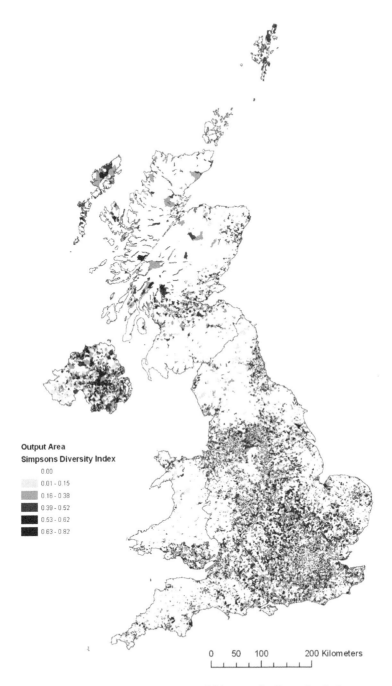

Figure 6.5   UK spatial distribution of Simpson's diversity index

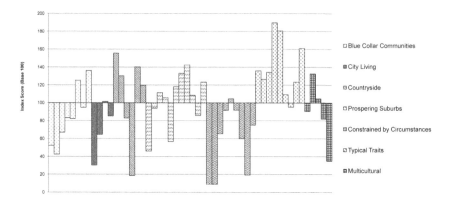

Figure 6.6   Incidence of high Simpson's diversity index scores by OAC Sub Groups

order to create clusters which represent a finer level of detail. This could possibly explain why the Groups "Village Life" and "Accessible Countryside" show greater heterogeneity. The Super Group "Typical Traits" represent some of the most heterogeneous areas with the exception of one Group, "Young Families in Terraced Homes" which presumably because of the spatial organisation of the housing stock, appears to have greater social similarity at the level of OA. In the Super Group "Multicultural", the "Afro Caribbean Communities" Groups are more homogeneous than the "Asian Communities" Group. Through examination of the Mosaic typology and pen portraits, it appears that the Mosaic classification places a great deal of emphasis on differentiating between wealthy and deprived Asian neighbourhoods. These neighbourhood Types appear across two Mosaic Groups, and therefore be associated with greater diversity within OAs.

A Total Weighted Deviation Evaluation of Course and Institutional Profiles

The performance of a classification to discriminate between groups within a target population may be measured by creating a total weighted deviation score (TWD). This was initially proposed by Webber (2004) as a method of evaluating geodemographic classifications relative to a range of other composite indicators. However, one flaw with this technique as conceived by Webber is the non linear scaling of index scores, i.e. scores can occur at much higher values above the average (100) than below the average. Therefore, an alternative measure of relative concentration is needed to compare a target and base population using a linear scale; that is, where categories which are overrepresented can appear to balance values which are underrepresented.

The following equations are derived from the index based method proposed by Webber (2004). Equation 6.3 shows that for a discrete classification, the total base population is divided into b_n categories The counts are converted into a proportion p where $_i$ denotes the discrete classification category from 1 through to n.

Equation (6.3) –

$$p_i = \frac{b_i}{\sum\limits_{i=1}^{n} b_i}$$

Equation 6.4 calculates a q_{ij} predicted rate for an i category by taking the n total postcodes recorded for each j variable (e.g. course grouping, ethnicity etc.) and multiplying by p_i across $p_1, p_2..., p_n$.

Equation (6.4) –

$$q_{ij} = n_j \times p_i$$

All q_{ij} predicted scores can then be substituted from the t_{ij} actual recorded frequencies to give the difference between the predicted and the actual counts as r_{ij}.

Once a set of r_{ij} scores are created for j_n variables across a discrete classification a total deviation (TD) statistic can be derived. A standardised statistic can be calculated across the differences (r_{ij}) to measure the dispersal of scores from the mean. The mean of this standardised statistic is 0 for all variables because the discrete classifications categorise 100% of the population, and where a discrete classification does not have a category for a unit postcode these are classified as "Unknown". Thus, the calculation to derive TD where p is the frequency of categories for the discrete classification is shown in Equation 6.5. The (p-1) correction in the denominator is not used as the calculation refers to the total population and not a sample (Hardy and Bryman, 2004).

Equation (6.5) –

$$TD = \sqrt{\sum\limits_{i=1}^{n} \frac{r_{ij}^{2}}{p}}$$

Total deviation statistics calculated for large populations should be considered more reliable as they are less vulnerable to "freak" patterns caused by outlier scores. Population weighting circumnavigates this potential problem, so the TD scores are multiplied by a proportion created by dividing the sum of the target variable frequency by the total base frequency (See Equation 6.6).

Equation (6.6) –

$$TWD = TD \times \frac{\sum\limits_{i=1}^{n} t_{ij}}{\sum\limits_{i=1}^{n} b_i}$$

TWD is thus a measure of classification performance on a linear scale which accounts for a variable's population size in its derivation. The technique is suitable for comparison of variable sets across a number of classifications, and thus provides a method of evaluating classification performance where a larger TWD is attributable with better discrimination between areas. Thus, for the comparison in this chapter, these TWD measures of predicted versus observed postcode frequency were calculated from UCAS 2004 acceptance data for courses and institutional profiles. The UCAS data were aggregated by the following classifications:

- Mosaic Groups (MGroup)
- Mosaic Types (MType)
- ACORN Category (ACat)
- ACORN Group (AGroup)
- ACORN Type (AType)
- POLAR (POLAR)
- Index of Multiple Deprivation (IMD)
- National Statistics Socio Economic Classification (NS-SEC)
- OAC Group (GROUP)
- OAC Sub Group (SUB)
- OAC Super Group (SUPER)
- Educational OAC Groups (EDUOACG)
- Educational OAC Types (EDUOACT)

The TWD scores are presented in Table 6.5; and in Figure 6.7 these scores are plotted against the frequency of clusters in each classification. It can be seen that there is a positive linear relationship between cluster frequency and TWD score, indicating that an increase in clusters results in a classification which expresses greater explanatory power.

Figure 6.7 is useful as it begins to explain how well a classification is performing relative to the frequency of clusters. Although the position of the fitted line will vary marginally as classifications are removed or added to the evaluation sample, those classifications which appear above the line (i.e. increasing TWD) could be interpreted as candidate parsimonious models. ACORN Groups, ACORN Types and Educational OAC Groups appear marginally above the line; however the explanatory powers (TWD) of these classifications vary widely within this range. ACORN Types is the highest performing classification, although Mosaic Types, OAC Sub Groups and Educational OAC Types also perform very well. OAC Sub Groups and Educational OAC Types appear to perform quite similarly. It would have been desirable if the performance of Educational OAC Types had increased by a greater proportion over OAC Sub Groups, however, as the build methodology was directly linked to OAC, this may have had a bearing on the performance gain. The increased performance experienced by the unit postcode level classifications of ACORN Types and Mosaic Types over those which classify at OA could be related to these classifications using a finer geographical scale. Thus, there may

Table 6.5   Total weighted deviation results

	Mtype	Mgroup	POLAR	IMD	NS-SEC	Group	Sub	Super	EduOACT	EduOACG	AGroup	ACat	AType
Institutions	27.2	7.3	4.7	8.1	5.4	11.3	22.9	5.0	22.7	7.2	11.2	4.7	27.3
Courses	26.6	8.0	4.3	8.7	5.0	10.4	22.6	5.0	24.6	7.7	11.1	5.0	27.9
SUM	53.8	15.2	9.0	16.8	10.4	21.7	45.5	10.0	47.4	14.9	22.3	9.7	55.2
Clusters	61	11	5	10	6	21	52	7	53	10	17	5	56

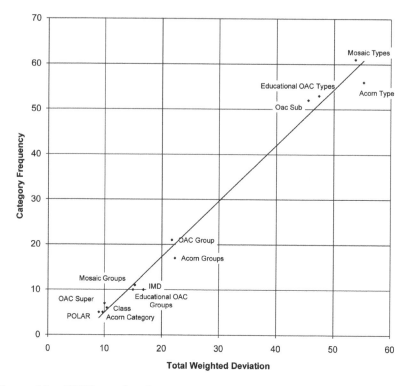

**Figure 6.7   TWD results plotted against the frequency of clusters
in the classification**

be an argument to suggest that an Educational classification should be built at the
scale of unit postcode. However, at Group level, this advantage appears eroded
with Educational OAC Groups only marginally underperforming Mosaic Groups.

A Lorenz and Gini Coeffient Evaluation of Young Participation in Higher Education

A further evaluation was conducted using a Lorenz curve (Lorenz, 1905) which
illustrates the gain in discrimination one would expect by using a particular
classification over the null hypothesis (straight line) that all areas supply
participants to Higher Education in equal proportions. The larger the area under
the curve, the greater the total discrimination, where a theoretical maximum score
would be equal to one. These area scores were calculated using a Gini coefficient
(Gini, 1912) as shown in Equation (6.7) where x are the 18–19 year olds in cluster
k and y are the 18–19 year old participants in cluster k.

Equation (6.7) – $$G = \left| 1 - \sum_{k=1}^{n} (y_{k-1} + y_k)(x_{k-1} - x_k) \right|$$

The Gini coefficients are shown in Table 6.6 and Figure 6.8.

Table 6.6   Gini coefficients

Rank	Classification	Gini	Clusters
1	Mosaic Types	0.4004	61
2	ACORN Types	0.3504	62
3	Mosaic Groups	0.3497	11
4	Educational OAC Types	0.3353	53
5	OAC Sub Groups	0.3258	52
6	ACORN Group	0.3245	17
7	OAC Groups	0.3156	21
8	ACORN Category	0.2923	5
9	OAC Super Groups	0.2837	7
10	Educational OAC Groups	0.2550	10
11	IMD	0.2419	10

As described previously, there is a high correlation between participation in Higher Education and wealth. The income measures which are included as input variables to create Mosaic are not available in the 2001 Census, and therefore are not included as input variables in OAC, nor its derivatives such as Educational OAC. Mosaic Types and Mosaic Groups, the latter only having 11 clusters, both perform very well when targeting low participation areas, presumably because of the inclusion or weighting of these extra variables relating to affluence. ACORN Types, Groups and Category perform less well than one might expect for a commercial classification. The 53 cluster Educational OAC Type classification is shown to outperform the 52 cluster OAC Sub Groups, however the Educational OAC Group classification does not perform well in predicting aggregate participation rates.

There are several important findings which can be drawn from both the TWD and Gini coefficient analysis. Although it may be desirable for an aggregate participation rate to be predicted by a classification (as measured in the Gini Coeffcent analysis), this is a less useful application than, for example, predicting a particular course participation propensity (as measured in the TWD analysis). With the advent of national educational databases such as those accessed for use in this book, there is little need to predict aggregate participation when actual participation rate can be derived and mapped (e.g. POLAR). Participation is correlated with "wealth", and classifications that stratify neighbourhood Types by this attribute will demonstrate a stronger prediction of participation, as has been

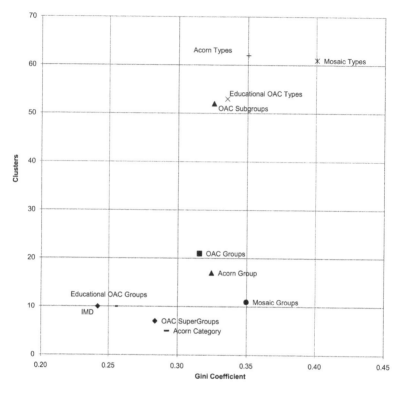

Figure 6.8   Gini coefficients and clusters

shown by the performance of the Mosaic classification in the Gini Coefficient analysis. However, the TWD analysis demonstrated that when disaggregating participation into course and institution categories, Mosaic performed less well than might be expected, being out performed by ACORN Types. One possible interpretation of this finding is that the Mosaic classification may stratify too highly by measures of affluence, and not enough by other attributes, which may influence or be important for measuring disaggregated applicant behaviours.

Conclusion

This chapter has shown how geodemographic classifications can be evaluated both quantitatively, for their relative discriminatory power, and qualitatively, to assess classification features against user requirements. Cross tabulations between classification typology have shown neighbourhoods possessing similar characteristics between the typologies. Through the comparison of OAC with Mosaic, it was also possible to see which Mosaic Groups are informed by data related to income, and which are absent from OAC. A detailed study of the relative information loss one can expect from switching between a classification that categorises neighbourhood at OA rather than unit postcode was examined. On the assumption that Mosaic represents an accurate depiction of reality, it was found that OA diversity was generally higher in urban rather than rural areas. Furthermore, the patterns of diversity were examined by OAC Sub Groups which indicated where the additional data used to construct Mosaic might provide extra discrimination, or, that through weighting schema, particular characteristics have been created in the Mosaic classification. A series of classification were compared using a total weighted deviation and Gini Coefficients. These measure the ability for classification to account for variations in participation rates, and disaggregated participation behaviours at course and institution level. Possible issues with the way in which the Mosaic typology stratifies neighbourhoods were highlighted in this analysis.

Chapter 7
Towards a More Meritocratic Market?

Introducing Temporal Access Change

Early geodemographic classifications were updated relatively infrequently using predominantly decennial census data. However, contemporary commercial classifications are privy to greater sources of data which are refreshed more frequently. Similarly, in the public sector data are collected annually, although unlike many of the commercial datasets which often relate to small samples of the total population, the public sector data are often far more comprehensive in target population coverage. The rate of change in the Higher Education sector in particular is very rapid, and therefore it makes sense to consider the longitudinal profile of the recent past.

Measuring if Current "Widening Participation" Policy is Successful

The UK government set a target of attaining 50% of young people aged below 30 to take part in Higher Education (DfES, 1999) which relates to widening of absolute participation but does not address access inequalities that could be either created or sustained by this growth. Although this secondary aim was not initially highlighted, the government have since re-stated their ambition that this growth should also extend access to underrepresented groups, highlighted by Hodge (2002b):

> We could get there without changing in any way the sort of people that go to university. If we do that, I will feel we will have failed. (Hodge, 2002b)

This chapter will not address the extent that the government are meeting the overall 50% participation target, but will instead focus on temporal changes in access by those groups underrepresented in Higher Education, focusing specifically on those differences between institutions and courses of study. As was discussed in Chapter 4, institutional performance for extending access was previously measured through a number of different performance indicators which for young participants include state school entrants, the National Statistics Socio-Economic Classification (NS-SEC) categories 4–7 and a measure of neighbourhood participation disadvantage using Super Profiles. These university performance indicators are assessed against a benchmark target, calculated using methods described in Draper and Gittoes (2004). Chapter 4 has demonstrated that these current definitions of widening participation students are in some sense problematic when compared to more modern area classification methods. HEFCE commissioned a review of performance indicators in August 2006 which was published in June 2007 (HEFCE, 2007a). In light of this review, the POLAR classification was adopted as an indicator of neighbourhood level HE participation disadvantage. It was also concluded that Geodemographic classifications, would disappointingly no longer form part of the performance indicator.

The use of official performance indicators to evaluate changes in access rates over time is limited. This limit is discussed by the Higher Education Statistics Agency (HESA) who notes that performance indicators are not designed to measure an individual institution against the sector, and as such they are not appropriate to construct a hypothetical performance league table:

> No meaningful league table could fairly demonstrate the performance of all higher education institutions relative to each other. The HE sector is extremely diverse. Each institution has its own distinct mission, and each emphasises different aspects of higher education. Because of this diversity, and the need to compare HEIs fairly, we have used a range of indicators and benchmarks. (HESA, 2004)

Thus, those institutional benchmarks created by HESA control for a series of institutional differences, including student's subjects of study, the qualifications held on entry and age on entry. Furthermore, some benchmarks are adjusted by the region of the country where the student's home is located. HESA (2004) state that the benchmarks can be used in two ways:

- To see how well an HEI is performing compared to the HE sector as a whole.
- To decide whether to compare two institutions.

The first point relates to how one can compare institutional performance above or below a benchmark against the same indicator in another institution, rather than comparing the actual profiles as raw counts, percentages or index scores. The second point concerns how meaningful it is to compare institutions, and by using benchmarks, how one can gauge the similarity between two institutions, and qualitatively assess whether it is appropriate to compare them. Higher Education benchmarks are therefore designed to standardise structural differences between institutions which, although possibly useful in terms of funding allocation, hides the underlying inequality that exists within the sector. Goldstein and Spiegelhalter (1996:385) describe performance indicators as "a summary statistical measurement on an institution or system which is intended to be related to the 'quality' of its functioning", where in the case of the HESA performance indicator, "quality" is held as a relative and standardised measure.

Perhaps a more appropriate method of evaluating changes over time could be a refined version of league tables, like those composed each year by many of the national press (E.g. *The Times Good University Guide,*[1] *Guardian University Guide*[2]). League tables take a different view on "quality", and one in which institutions are compared and ranked by their merits or failures across a broad range of assessment criteria and dislocated from any contextual factors. It is usual for national averages to be used for these comparisons, as the overall aim is typically to create a ranking above or below this point. In the context of extending access, these types of measures can be used to demonstrate which institution or course type is:

- Increasing the frequency of admitted students from low participation backgrounds.
- Increasing the frequency of admitted students from low participation backgrounds relative to institutional growth.
- Increasing the frequency of admitted students from low participation backgrounds relative to institutional growth and changes in the background frequency of these groups across society.

Therefore, using various UCAS data from 2001–2006, this chapter aims to explore both absolute trends in participation and widening access and also disaggregated trends within institutions and course groupings. The temporal period considered within each analysis varies, depending on the availability of data or variables for a particular period in time. Adjusted benchmarks will therefore not be used in these comparisons, as the aim is to assess the absolute, rather than standardised (relative) differences, and to demonstrate the true stratification of access over time, rather than measure standardised performance at improving access inequality.

As discussed in Chapter 2, there are multiple definitions of Higher Education, and one way of demonstrating that participation has improved is to increase the definition of which courses of study (and therefore people) are included within "Higher Education". The 50% national participation target discussed above is based on students aged between 18 and 30, and includes any course which is above an A-Level (i.e. Level 4+ in the National Qualifications Framework) and leads to a qualification which is awarded by a Higher Education institution or national awarding body (Prospects, 2002). This is a broad definition of Higher Education and is not adopted in the benchmarks produced by HESA, who manage statistics for a more compact range of institutions. The analysis presented in this chapter concerns those who have accepted full time degree courses at UCAS institutions.

---

1   www.thegooduniversityguide.org.uk.
2   www.education.guardian.co.uk/universityguide/0,,488282,00.html.

Accounting for Higher Education Growth

As shown in Chapter 2, the Higher Education sector has continued to expand and increasing participation towards a 50% participation target has continued to occur. However, the extent that this growth has occurred evenly across all subjects can be investigated using UCAS data from 2002–2006 (see Table 7.1). This table shows that there are course groups which have grown by as much as 39.8% (D Veterinary Science, Agriculture and Related) whereas others have declined by 24.7% (Non-European Languages and Related). There are too many subjects to analyse individually, so a number of the subject groups are examined in more detail at Line level over the following sections.

Courses in Decline – Group G: Mathematical and Computer Science

Students within Group G declined by 14.6% between 2002 and 2006, amounting to 3,947 fewer degree acceptances. However, not all subject Lines are declining within this Group (see Figure 7.1).

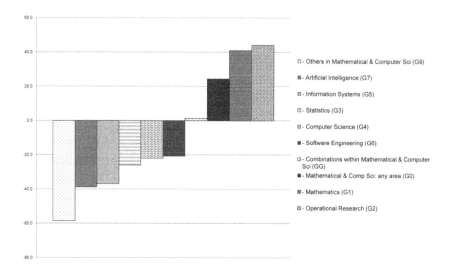

Figure 7.1   The percentage decline in subject line popularity within Group G: Mathematical and Computer Science

There is a clear dichotomy within Group G between the mathematical growth subjects and the computational decline subjects. In 2002, Computer Science received the largest frequency of all degree acceptances; however, by 2006, this had fallen to sixth place behind Design Studies, Law, Psychology, Management and Business Studies (See Table 7.2). One surprise entry in this list is Social

Table 7.1 JACS course popularity by degree acceptances ranked by the percentage change in these frequencies between 2002 and 2006

Subject (JACS Group)	2002	2003	2004	2005	2006	% Change
T Non-European Languages and Related	2,245	2,162	2,113	1,935	1,691	-24.7
General, Other Combined and Unknown	6,714	6,439	5,590	5,871	5,519	-17.8
G Mathematical and Computer Science	26,978	25,597	23,273	23,886	23,031	-14.6
Combined Social Sciences	5,189	5,381	5,123	5,123	4,903	-5.5
H Engineering	20,740	20,659	21,046	21,339	20,434	-1.5
R European Languages, Literature and Related	4,010	3,841	3,889	4,220	4,023	0.3
Combined Arts	11,706	11,544	11,354	12,500	11,813	0.9
Social Sciences Combined with Arts	10,126	10,241	10,102	10,711	10,333	2.0
V Historical & Philosophical Studies	12,685	12,535	12,966	14,096	12,985	2.4
Sciences Combined with Social Sciences or Arts	17,022	17,234	16,549	18,327	17,542	3.1
F Physical Sciences	14,145	1,4305	13,878	14,980	14,927	5.5
J Technologies	2,290	2,190	2,334	2,314	2,418	5.6
Q Linguistics, Classics and Related	10,529	10,744	10,991	11,855	11,324	7.6
Combined Sciences	5,811	5,404	5,410	6,203	6,334	9.0
P Mass Communications and Documentation	8,115	8,588	8,720	9,770	9,063	11.7
C Biological Sciences	27,657	28,982	29,262	32,446	30,916	11.8
N Business and Administration Studies	39,235	41,006	42,048	45,318	44,843	14.3
M Law	15,921	17,733	18,580	19,084	18,448	15.9
A Medicine and Dentistry	7,885	8,615	8,944	9,008	9,146	16.0
L Social Studies	23,983	24,575	26,400	29,980	29,344	22.4
W Creative Arts and Design	33,233	35,505	38,198	42,447	40,924	23.1
X Education	10,003	10,819	11,843	13,138	13,082	30.8
B Subjects Allied to Medicine	19,660	21,702	23,744	26,416	26,496	34.8
K Architecture, Building and Planning	5,855	6,560	7,273	8,380	8,128	38.8
D Veterinary Science, Agriculture and Related	2,844	3,170	3,355	3,969	3,976	39.8
Total	**344,581**	**355,531**	**362,985**	**393,316**	**381,643**	

Work, rising 39 places. This is a slightly artificial rise in popularity and is accounted for by the merger of the Social Work Applications System (SWAS) with UCAS in 2004.

Table 7.2   Top 10 most popular degrees in 2006 and their ranked position change since 2002

Rank	2006 Top 10 Degree Subjects	Degree Acceptances	Rank Change
1	Design Studies	15,193	+1
2	Law by Area	14,982	+1
3	Psychology	13,113	+2
4	Management Studies	12,142	0
5	Business Studies	11,143	+1
6	Computer Science	10,987	-5
7	English Studies	8,862	0
8	Sports Science	8,598	+7
9	Social Work	8,303	+39
10	Pre-clinical Medicine	8,011	+1

Mander (2007) suggests that these patterns may be attributable to an eroding advantage perceived by the general public for obtaining a computational degree since the 2001 dot-com bubble burst. This may be true, although the reality is likely much more complex, with many other contributing factors, including the rising use of high level application development environments within the private sector, a shifting commercial focus away from bespoke and low level development into application led solutions technologies, such as those provided by the likes of SAP[3] or Oracle[4] and the fragmentation of Computer Science into other disciplines such as GIS in Geography. Within this context, this decline could be understood as a market re-adjustment to over supply against changes in demand.

Courses with Little Growth – Group F: Physical Sciences

Although Group F Physical Sciences showed little growth, like subject Group G, there is variability in course Line popularity (See Table 7.3). There was a huge growth in Forensic and Archaeological Science of 57.5% which increased student degree acceptances by 674 places. Sir Howard Newby noted to a committee of MPs looking at declining applications to science degrees:

---

3 www.sap.com.
4 www.oracle.com.

There has been a big drive towards forensic chemistry, thanks to Amanda Burton[5]...I'm not quite sure who is going to employ all those forensic scientists. (Newby, 2005)

Although there has been much high profile interest in the closure of specific Chemistry departments such as the University of Sussex,[6] Exeter[7] and threats of closure at Kings College London,[8] this does not seem to have affected overall growth as Chemistry grew by 17.7% with around 538 extra places.

Table 7.3 **JACS course line popularity within Group F Physical Sciences by degree acceptances and ranked by the percentage change in these frequencies between 2002 and 2006**

	2002	2003	2004	2005	2006	% Change	Frq Change
Others in Physical Sciences	1,098	988	775	801	734	-33.2	-364
Astronomy	196	163	170	142	146	-25.5	-50
Materials Science	4	3	13	8	3	-25.0	-1
Combinations within Physical Sciences	621	603	522	556	528	-15.0	-93
Geology	1,389	1,402	1,344	1,304	1,258	-9.4	-131
Physics	2,907	2,905	2,671	2,927	2,934	0.9	27
Physical and Terrestrial Geography and Environmental Sciences	3,321	3,286	3,149	3,517	3,464	4.3	143
Physical Sciences: any area of study	202	238	215	227	221	9.4	19
Ocean Sciences	189	199	214	210	209	10.6	20
Chemistry	3,045	3,042	3,080	3,464	3,583	17.7	538
Forensic and Archaeological Science	1,173	1,476	1,725	1,824	1,847	57.5	674

5   Who is the lead character in the popular BBC forensic drama "Silent Witness".
6   www.news.bbc.co.uk/2/hi/uk_news/england/southern_counties/4847000.stm.
7   www.news.bbc.co.uk/2/hi/uk_news/education/4114919.stm.
8   www.news.bbc.co.uk/2/hi/uk_news/england/london/2944131.stm.

Courses in Growth – Group L: Social Studies

Group L Social Studies grew by 22.4% between 2002 and 2006 with 5361 additional degree acceptances. Much of this growth is attributable to the additional courses included within the Line Social Work after the agglomeration of the Social Work Admissions Systems (SWAS) into UCAS (See Table 7.4). Whereas subject Lines within this Group, such as Politics, have seen an 18.7% rise in degree acceptances, others, such as Human and Social Geography, have fallen by 16.9%, with around 415 fewer degree acceptances in 2006 than in 2004.

Table 7.4   JACS course line popularity within Group L Social Studies by degree acceptances and ranked by the percentage change in these frequencies between 2002 and 2006

	2002	2003	2004	2005	2006	% Change	Frq Change
Social Studies: any area of study	754	723	707	697	631	-19.5	-123
Human and Social Geography	2,870	2,738	2,722	2,739	2,455	-16.9	-415
Anthropology	697	571	534	618	619	-12.6	-78
Sociology	4,577	4,320	4,218	4,682	4,116	-11.2	-461
Combinations within Social Studies	2,554	2,731	2,436	2,607	2,461	-3.8	-93
Economics	5,502	5,524	5,498	5,445	5,314	-3.5	-188
Social Policy	897	813	856	920	952	5.8	55
Politics	3,375	3,539	3,817	4,334	4,152	18.7	777
Others in Social Studies	253	244	282	281	341	25.8	88
Social Work	2,504	3,372	5,330	7,657	8,303	69.8	5,99

Widening Participation Profiles Over Time

In Chapter 2, it was shown that changes in the proportion of students attending Higher Education from different social class groups had remained relatively static since 1968. However, as was discussed in Chapter 3, these measures are increasingly unreliable because of the fluidity of the current job market and the assignment of the groups based on parental occupation. It was then argued that geodemographic analysis provides a more robust measure of socio-spatial stratification. However, before temporal rates of neighbourhood access are considered, a point raised in Chapter 2 will be addressed. This was an increasing trend for accepted applicants to be recorded by UCAS with a current occupation classification of "unknown".

Social Class and the "Unknowns"

One measure of increasing access for Higher Education institutions is the frequency of students who accept places on their degrees by their National Statistics Socio Economic Classification (NS-SEC). The frequency scores were obtained from the UCAS website for 2002–2006 and are shown in Figure 7.2 as, prior to 2002, a different socio-economic classification was used. In terms of absolute numbers, there has been little change in the frequency of students between groups; however there has been a growing trend for the increase in students classified as "unknown".

In order to look at the relative access rates between years a measure of the change in base employment characteristics over the period 2002 to 2006 was required as this has an effect on the supply of different people into Higher Education from backgrounds classified into NS-SEC. The largest sample data available were found in extracts from the National Statistics Labour Force Survey (LFS).[9] The 2002–2006 base data were combined with degree acceptances counts over the same time period, and index scores were calculated (see Figure 7.3).

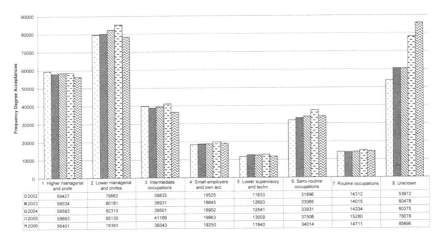

	1. Higher managerial and profs	2. Lower managerial and profes	3. Intermediate occupations	4. Small employers and own acc	5. Lower supervisory and techn	6. Semi-routine occupations	7. Routine occupations	8. Unknown
2002	59427	79662	39833	18525	11833	31896	14312	53972
2003	58034	80181	38931	18845	12693	33065	14015	60478
2004	58593	82310	39501	18952	12541	33931	14334	60375
2005	58693	85130	41189	19963	13009	37506	15280	78078
2006	56401	78383	36543	19250	11940	34014	14711	85696

Figure 7.2   Frequency of degree acceptances – 2002–2006 by NS-SEC

From this analysis, it appears that access is marginally improving for NS-SEC 6 and 7, with little change for 4 and 6. The NS-SEC category which is most in relative and absolute decline is the most affluent group 1, which is an interesting finding, since one might expect these groups to be either growing or maintaining their overrepresentation. The larger relative change may, however, be caused by the census base count age range not matching the target count exactly. Those students from the UCAS data classified as "unknown" have been excluded from

---

9   www.statistics.gov.uk/STATBASE/Source.asp?vlnk=358.

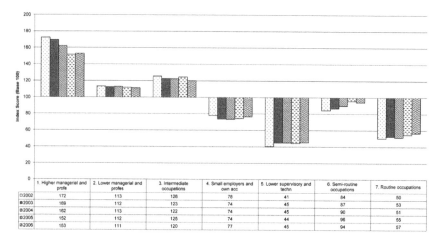

Figure 7.3   Index scores of degree acceptances – 2002–2006 by NS-SEC

this analysis as they cannot be compared to a base distribution, as all results from the LFS sample are categorised within one of the seven NS-SEC.

Using 2004 UCAS data, the "unknown" classified students were cross tabulated against a series of variables in order to understand why and where these patterns were occurring. The first hypothesis was that certain types of schools may be deliberately advising students not to fill in these details as a method of manipulating any disadvantage that a high income family may incur during the application process, or, that certain school types were not providing the support to enable students to fill in forms correctly. This was tested by examining the distribution of "unknown" students within by the UCAS classification of schools (See Table 7.5).

Table 7.5   The percentage of 2004 degree acceptances classified as "unknown" by UCAS School Type

School Type	Percentage
Six Centres and Colleges	13%
Comprehensive School	10%
Grammar School	9%
Independent School	9%
FE/HE	25%
Other	34%

Around 10% students within each school category are coded as unknown, with the exception of FE/HE and Other. Other institutions are difficult to assess, as they are not classified into any particular category, however, FE/HE colleges are larger institutions. Universities appear in the UCAS school classification because some applicants apply for course through UCAS once they have already begun a different course, e.g. a switch from Sociology to Geography. If the "UCAS schools" containing the 20 highest frequency of students coded as unknown are examined, this confirms these findings (see Table 7.6). It should also be noted, the "Newham VI Form College" duplication is not a mistake, as the UCAS schools data contains many duplicate records. These duplicates often refer to separate campuses, and as such are identified by a separate UCAS school code.

Table 7.6   2004 degree acceptance classified as "unknown" by UCAS School Type

Name	Unknown	All ACC	%
University of the Arts London	485	1,327	37
Westminster Kingsway College	234	452	52
Newcastle College	227	841	27
Barnet College	222	626	35
Tower Hamlets College	214	361	59
City and Islington College	201	357	56
Newham VI Form College	191	441	43
Leyton VI Form College	180	390	46
Luton VI Form College	159	542	29
Uxbridge College	158	395	40
Methodist College	157	202	78
Sir George Monoux College	154	418	37
Wigan and Leigh College	153	254	60
Croydon College	151	413	37
Richmond Upon Thames College	147	875	17
East Lancashire Institution of He	141	491	29
City of Westminster College	138	331	42
Sutton Coldfield College	136	576	24
NW Institute of F and H Education	133	303	44
Reading Coll and School of Art/Design	133	388	34

For universities that are re-admitting students through UCAS (e.g. University of the Arts London who have 37% of their successful degree applicants classified as unknown) it could have a potentially negative bearing on the widening participation funding they are allocated. Only those students who have NS-SEC classifications are included in the benchmarks calculated by HESA, and as was shown in Chapter 2, most of this information is derived from UCAS data, i.e. the institutions rarely add to the data they receive from UCAS at the end of the application cycle. Thus, if students from low NS-SEC 4–7 had been classified as "unknown" during the application process, the institutional performance against their benchmark score would appear lower than it is in reality.

Table 7.7 demonstrates that there also appears to be an ethnic dimension to the classification of "unknown" students, with Asian-Bangladeshi, Asian-Pakistani and Black-African all having a large proportion of accepted applicants coded as "unknown". This may be a cultural factor, in that jobs typically conducted by the parents of these students are difficult to classify (perhaps self employment), or that these students on aggregate attend those schools which provide less support when completing UCAS applications.

Table 7.7   **2004 degree acceptances classified as "unknown" by ethnic group**

Ethnic Classification	Acceptances	Unknown	%
Asian – Bangladeshi	2,681	1,261	47.0
Asian – Pakistani	7,999	3,105	38.8
Black – African	8,076	3,041	37.7
Black – Other black background	843	276	32.7
Other ethnic background	2,661	830	31.2
Asian – Other Asian background (ex. Chinese)	3,427	914	26.7
Black – Caribbean	3,584	903	25.2
Mixed – White and Black African	772	187	24.2
Chinese	3,311	761	23.0
Asian – Indian	13,147	2,859	21.7
Mixed – White and Black Caribbean	1,611	337	20.9
Mixed – Other mixed background	2,378	469	19.7
White – Other white background	8,764	1,626	18.6
White – Irish	10,136	1,538	15.2
Mixed – White and Asian	2,456	371	15.1
White – Scottish	22,356	3,089	13.8
White – Welsh	10,902	1,418	13.0
White – English	195,834	25,230	12.9

To gauge the neighbourhood stratification of these students, the ACORN geodemographic profile for the "unknown" category is shown for 2004 in Figure 7.4. The neighbourhoods which are overrepresented are from both affluent and less affluent urban areas.

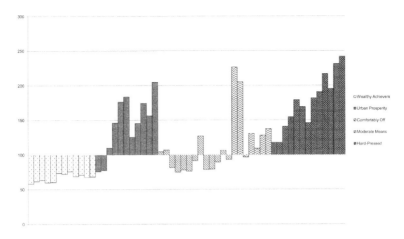

Figure 7.4   2004 degree acceptance classified as "unknown" by ACORN Types

Are Neighbourhood Access Rates Improving?

The ACORN Group profile from 2001–2004 for degree acceptances is shown in Figure 7.5. There does appear to be a marginal shift in the participation propensity between those more affluent neighbourhoods (left side of the chart) towards those less affluent neighbourhoods (right side of the chart), therefore suggesting that on aggregate, widening participation initiatives are working to extend access and increase the representation of underrepresented neighbourhood groups. However,

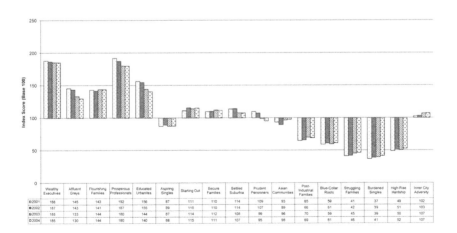

	Wealthy Executives	Affluent Greys	Flourishing Families	Prosperous Professionals	Educated Urbanites	Aspiring Singles	Starting Out	Secure Families	Settled Suburbia	Prudent Pensioners	Asian Communities	Post-Industrial Families	Blue-Collar Roots	Struggling Families	Burdened Singles	High-Rise Hardship	Inner City Adversity
2001	188	145	143	192	156	87	111	110	114	109	93	85	58	41	37	49	102
2002	187	143	141	187	155	99	115	110	114	107	89	66	61	42	39	51	103
2003	185	133	144	180	144	87	114	112	108	98	96	70	59	45	39	50	107
2004	185	130	144	180	140	88	115	111	107	95	98	69	61	46	41	52	107

Figure 7.5   2001–2004 degree acceptance index scores by ACORN Groups

this does not include the period from 2006 onwards, which was the first year "top-up" fees were introduced. The distribution of NS-SEC shown earlier highlighted that over this period there was not a large change in the distribution as many predicted (see Galindo-Rueda et al., 2004), and therefore one may expect a similar neighbourhood profile.

Which Subject Areas are Extending Participation the Most?

The following analysis extends from the aggregate change in neighbourhood participation rates presented in the previous section to examine all subjects in relation to their changing geodemographic profiles in order to assess which are most effective in extending access. For the purposes of this analysis, widening participation students are classified as those from neighbourhoods identified by the ACORN Category "Hard Pressed", which represents the Category which is least affluent and has the lowest participation rate in Higher Education. There are several methods that may be adapted to assess widening access, and the appropriateness of the measures is to some extent determined by the aims of the analysis. For example, those interested in aggregate growth in widening participation students may only be interested in those courses which have recruited the highest absolute number of widening participation students. However, a more appropriate measure for identifying the courses that are extending access would entail comparison of the differences between the index scores of the students admitted in 2002 compared to 2004, as these control the relative growth and contraction of the total course sizes, and also variability in the ACORN Categories in the underlying total population. Therefore, this analysis demonstrates:

- The change in total number of widening participation students admitted to a course between 2002 and 2004.
- The index change in the total frequency of widening participation students admitted to a course between 2002 and 2004.

Table 7.8 shows the top and bottom courses ranked by their absolute change in widening participation student frequency. These scores reflect changes in the total course sizes, and also the courses which have contributed most or least in absolute terms to aggregate extending access profiles. Of those courses where student total numbers are in decline, that were discussed earlier (e.g. Group G), these also have contributed most to the decline in widening participation students. Thus, a course which typically attracts only a small proportion of widening participation students would lose only a small absolute number of students if the overall frequency of students declined.

Table 7.9 shows the ranked index differences for widening participation students in 2002 when compared to 2004. The difference between these two figures is the change in how over- or under- represented the widening participation students are, in comparison to the total population of students when controlling for changes

Table 7.8   Top and bottom 10 courses for extending absolute access

Rank	Bottom 10	Frq Change	Top 10	Frq Change
1	G4 – Computer Science	-136	L5 – Social Work	713
2	G5 – Information Systems	-127	M1 – Law by Area	457
3	H6 – Electronic and Electrical Engineering	-70	B7 – Nursing	303
4	N2 – Management Studies	-45	W2 – Design Studies	235
5	C1 – Biology	-39	B9 – Others in Subjects allied to Medicine	153
6	C7 – Molecular Biology, Biophysics and Biochem.	-38	C8 – Psychology	143
7	F3 – Physics	-34	C6 – Sports Science	142
8	L3 – Sociology	-33	A1 – Pre-clinical Medicine	123
9	F8 – Physical and Terrestrial Geog and Env Sci.	-31	N4 – Accounting	114
10	N6 – Human Resource Management	-27	W6 – Cinematics and Photography	98

Table 7.9   Top and bottom 10 courses for extending access by index change in widening participation students where total students in 2004 > 400

Rank	Bottom 10	Index Change	Top 10	Index Change
1	X3 – Academic Studies in Education	-6.2	G4 – Computer Science	15.9
2	C7 – Molecular Biology, Biophysics and Biochemistry	-4.8	F4 – Forensic and Archaeological Science	12.6
3	J9 – Others in Technology	-4.4	GG – Combinations within Mathematical and Computer Science	12.0
4	L7 – Human and Social Geography	-3.3	G5 – Information Systems	11.9
5	B8 – Medical Technology	-3.2	M9 – Others in Law	11.8
6	D3 – Animal Science	-3.1	N8 – Tourism, Transport and Travel	11.3
7	F8 – Physical and Terrestrial Geog and Environmental Science	-2.9	B6 – Aural and Oral Sciences	11.2
8	K2 – Building	-2.8	P9 – Others in Mass Communications and Documentation	9.9
9	F3 – Physics	-1.7	G6 – Software Engineering	9.8
10	Q8 – Classical Studies	-1.1	V3 – History by Topic	9.5

in these neighbourhood Category between the two time periods. The calculation entails subtracting the index scores for 2004 from those in 2002. Because courses with low total numbers of participants can produce volatile index scores, those courses with 2002 and 2004 total uptake of fewer than 400 students were excluded. The 400 student cut-off point meant that 44.7% of all courses were excluded from the analysis. The 55.3% of courses which remained accounted for 97.7% of the total undergraduate student population. A very interesting result to come from this analysis is the appearance of Computer Science at the top of the table for extending access, and other subjects from the same JACS group within the top ten. However, an alternative interpretation may be that because this course is in decline so rapidly, it is the non widening participation students who are choosing not to study Computer Science at a faster rate than the widening participation students (see Figure 7.6). It is highly unlikely that, across the UK as a whole, any course is actively and consistently choosing to recruit students from non widening participation backgrounds over widening participation backgrounds. There are, however, some courses which need to examine why they are attracting marginally fewer of these students in 2002 than 2004.

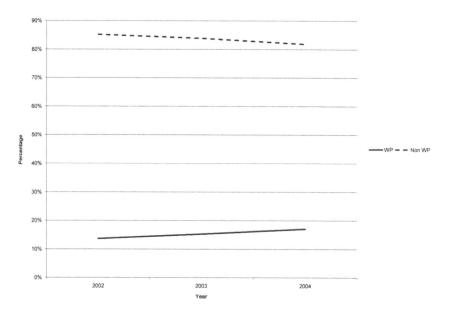

Figure 7.6   **The proportional change in the frequency of widening participation and non widening participation students between 2002 and 2004 in Computer Science (G4) degree recruitment**

Access and Participation Policy Interventions

Thus far the discussion on access has related to those general interventions by
HEFCE to help widen access through provision of funds linked to institutional
participation profiles. However, as discussed previously, some areas of Higher
Education are struggling not only to recruit widening participation students, but
also to maintain current levels of subject popularity. This was highlighted by
the example of the HEFCE "Strategically Important and Vulnerable Subjects"[10]
initiative, which examined those subjects which were of national importance,
and has since invested in a £160m programme of interventions to boost their
popularity. The remaining part of this chapter will examine a further initiative
which was funded by the then DCSF which aimed to address a fall in popularity of
the subject of Geography. This decline should be particularly worrying for Human
Geographers, as Table 7.3 previously showed Physical and Terrestrial Geography
and Environmental Sciences to have grown by 4.3%. Once the Physical and
Human Geography are combined, and the proportions compared to the total degree
acceptances for years 2002–2006 the overall trend for Single Honours Geography
can be seen in Figure 7.7. Chemistry has also been included for context.

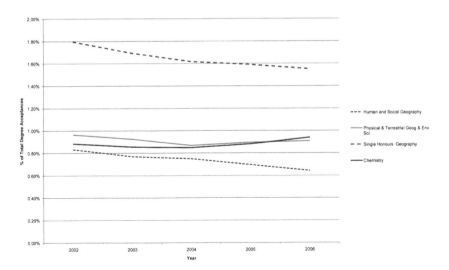

Figure 7.7   The proportion of degree acceptances in Geography

This is a worrying trend for Geography and also mirrors a decline in the course's
popularity at GCSE (See Table 7.10) and a critical report by Ofstead stating
that Geography was one of the worst taught subject in primary schools (Bell,

10   www.hefce.ac.uk/aboutus/sis.

2004). In 2006 a two million pound initiative was launched by the government
to address the decline (BBC, 2006). This "Action Plan for Geography" (RGS,
2006) consisted of:

- A new "Geography Teaching Today" (GTT) Website
- A Geography Ambassadors Programme
- Inspirational Geography Resources for Key Stage 3
- Fieldwork and Local Learning "Virtual Centre"
- Geography Leadership CPD
- Curriculum Development Projects
- Quality Marks and Chartered Geography Status

**Table 7.10   The frequency of 15 year old pupils in schools, and those
attempting a GCSE in Geography by end of the year**

	1999/ 2000	2000/ 2001	2001/ 2002	2002/ 2003	2003/ 2004	2004/ 2005	2005/ 2006
All	580,393	603,318	606,554	622,165	643,560	633,414	645,931
Geography	217,100	220,400	208,300.0	200,100	197,100.0	188,600	186,800

Sources: DCSF. There is no single source of these data so they were extracted from the
following documents present on the DFES website:
2005/2006: www.dfes.gov.uk/rsgateway/DB/SFR/s000702/index.shtml.
2004/2005: www.dfes.gov.uk/rsgateway/DB/SFR/s000664/index.shtml.
2003/2004: www.dfes.gov.uk/rsgateway/DB/SFR/s000585/index.shtml.
2002/2003: www.dfes.gov.uk/rsgateway/DB/SFR/s000442/contents.shtml.
2001/2002: www.dfes.gov.uk/rsgateway/DB/SFR/s000377/index.shtml.
2000/2001: www.dfes.gov.uk/rsgateway/DB/SBU/b000334/index.shtml.
1999/2000: www.dfes.gov.uk/rsgateway/DB/SBU/b000266/154.htm.

Although the Geography Action Plan contains worthy initiatives to increase quality
overall, the success of these monies has not been evaluated in terms of raising
absolute interest. Furthermore, if interest in Forensic Science in Higher Education
can be generated from the popularity of a range of television programmes such as
"Waking the Dead", "CSI" and "Silent Witness", then perhaps this initiative misses
an opportunity that has inadvertently proved successful for other disciplines. This
analogy is not directly applicable as Forensic Science does not appear on the
national curriculum in the UK; however nor does Geography in the US or many
other developed nations, where despite its unprivileged prior curriculum position,
there continues to be growing interest in the discipline. This is perhaps most
effectively demonstrated by the creation of the "Center for Geographical Analysis"
in 2006 at Harvard University (Gehrman, 2006). This centre focuses on GIS, which
Harvard University is calling the "new Geography". This direct comparison to the

US is perhaps a little unfair, as in the UK *new* Geography includes a plethora of study areas such as the geographies of car interiors (Ashton, 2005), otter hunting (Allen, 2005) and pet therapy (Kearns, 2005). Although these are a bias selection of topics covered in the talks from the RGS 2005 conference, they do highlight how fragmented the discipline has become in the UK. "Geography is the study of the Earth's landscapes, peoples, places and environments" (RGS, 2007), and it is argued here that this literal meaning should perhaps be replaced by something more appropriate that addresses key issues and problems at a scale where solutions or interventions can be effectively implemented.

The increased interest in Forensic Sciences generated by television programmes should provide evidence to the Geography advisory committee on how the student market responds to targeting initiatives, and although curriculum based initiatives may raise overall subject teaching quality, there is no guarantee that this will raise popularity. Enlisting the help of celebrity Geographers such as Michael Palin (now President of the RGS) for the launch event will help, although the Action Plan for Geography money may have been better spent on creating learning resources around television programmes with a geographical focus, or even enlisting promotional or marketing experts to reignite interest in the discipline amongst school children through initiatives which are demographically appropriate. The BBC partnership with the Open University is a very good example of how Geography education can be linked to television shows such as Coast through online[11] and offline course material and resources, and it is these activities which should be encouraged.

Perhaps what should be of more concern to Higher Education Geography departments than a declining popularity are the apparent increasing concentrations of students from more affluent areas in the overall student demographic. Between 2002 and 2004, the trends within Human and Social Geography indicate that the rate of loss from the subject is greatest from those students in less affluent neighbourhoods that are the focus of widening participation initiatives (see Figure 7.8). The main Groups which are increasing their over representation by around 20 index points are "Wealthy Executives" and "Prosperous Professionals". Those Groups also increasing at a slower rate of around 10 index points are "Secure Families" and "Asian Communities". The growth is predominantly in affluent middle class areas, where participation in Higher Education is more prevalent with the exception of "Asian Communities" where participation is slightly below average and in more deprived neighbourhoods.

There are clearly problems in both aggregate Human Geography recruitment and also the decline in degree acceptances from low participation neighbourhoods. The two million pound initiative aimed at early years education mentioned earlier will hopefully improve the situation by encouraging more students to study the subject; however, without Geography being on the core curriculum for secondary education, it will likely continue to be marginalised.

———————————

11   www.open2.net/coast.

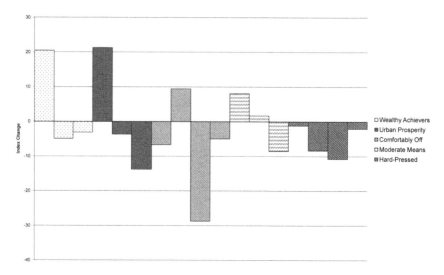

Figure 7.8   Change in Social Geography degree acceptance index scores between 2002 and 2004 by ACORN Groups

Conclusion

This chapter has examined how the Higher Education sector has variable growth rates between courses within institutions. Some courses over the period 2002–2006 have shown large increases, such as Forensic Science, whereas others are in rapid decline, such as Computer Science. The changing composition of the students who were attending both courses and institutions was examined, firstly through examining the NS-SEC classification which suggests a trend of increasing rates of students classified as of "unknown" NS-SEC. Students with missing data were found to be disproportionately from ethnic minority groups, within the FE sector or reapplying from within a university and also from predominantly urban areas. Overall access to Higher Education from low participation neighbourhoods improved marginally over the period 2001–2004, yet there still remained a large inequality in participation rates between those neighbourhoods where potential students were most and least likely to participate in Higher Education. Rates of change over this period were also shown to be heterogeneous between subject groupings.

Chapter 8

A Gallery of Applications
for Higher Education Stakeholders

Higher Education Stakeholders

Throughout this book, arguments have been built which suggest how a range of rich data pertaining to participation and performance in education can be integrated to provide both cross sectional and longitudinal information through relevant and bespoke indicators. For those stakeholders in Higher Education, that is students, schools and universities, these analyses help exemplify how more informed decisions can be made. Schools are seen as an important link in this chain as they contain those student "customers" of Higher Education, and thus schools need to be equipped to give suitable advice depending on students' needs and backgrounds. This chapter therefore presents a series of applications which utilise some of those techniques employed throughout this book and integrates them within broader Geographic Information Science techniques to investigate school "market areas".

A Regional Case Study – Stakeholders in Manchester

For this regional study, Higher Education stakeholder data is analysed to examine how patterns of socio-spatial differentiation in access to Higher Education are created in part as a function of local processes occurring in schools and colleges. The local authority of Manchester provides a useful analytical area as it contains a mix of school types with variable attainment.

Prior Attainment

Chapter 4 showed the general trends in pre- and post-16 neighbourhood level attainment without reference to local school effects. Students progressing into study at post-16, possess qualifications from a variety of schools, each with a different mix of students and attainment. The following analysis examines the interaction between neighbourhood attainment and school attainment in pre-16 schooling to demonstrate how prior attainment is spatially heterogeneous, thus creating an early filter on those students who will progress into Further and Higher Education. The average GCSE point scores were calculated for the neighbourhood groups within schools, and compared against the average school attainment, to analyse which neighbourhood groups within the schools were contributing most to the overall attainment. A further comparison was made by comparing the average attainment for neighbourhood Groups within a school against the same Groups in the whole of England to benchmark school performance in terms of raising the attainment for students from specific neighbourhood Groups. In this analysis, the attainment for neighbourhood Groups within a school were only analysed where there were more than 5 students within the Group to prevent average scores being taken from very small numbers of observations.

These comparisons are illustrated by two of the schools within the local authority of Manchester. The geodemographic profile of Parrs Wood High School for those completing GCSE at the end of KS4 is shown in Figure 8.1. Figure 8.2 shows how the attainment from these different neighbourhood Groups compared to the school average. Those students attaining the highest GCSE scores are those from the more affluent areas. In all of the neighbourhood level attainment graphs that are presented for schools in the following section, where the number of pupils within a group was five or fewer, then the results were suppressed. The numbers which label the columns are the frequencies of students within each neighbourhood group.

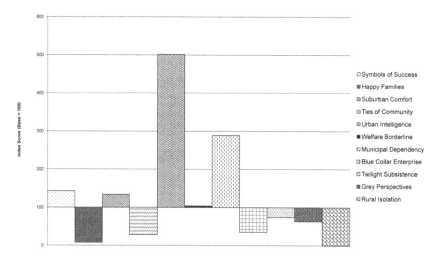

Figure 8.1   Mosaic profile of Parrs Wood High School – KS4

One can also examine how student GCSE performance varies according to residential neighbourhood Groups within the school, when compared to the national average, in order to assess how much better or worse the school is performing for students from within these areas than one might expect, based on the national average. Figure 8.3 demonstrates that this school is performing better than average across all Groups present in the school, with the exception of pupils from "Ties of Community" neighbourhoods, who perform slightly below average for their geodemographic profile.

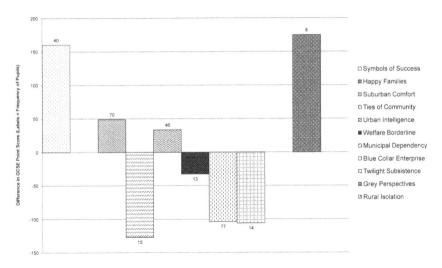

Figure 8.2   Average neighbourhood level GCSE point scores within Parrs Wood School compared with the school average GCSE score

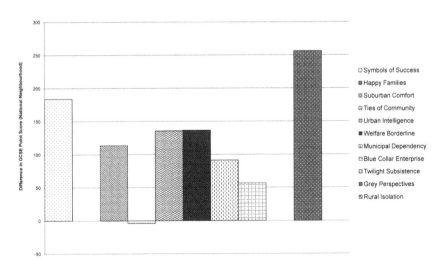

Figure 8.3   Average neighbourhood level GCSE point scores within Parrs Wood School compared with the national neighbourhood average GCSE score

The KS4 geodemographic neighbourhood profile for Whalley Range High School is shown in Figure 8.4.

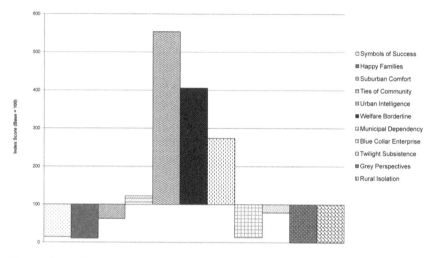

Figure 8.4   Mosaic profile of Whalley Range High School – KS4

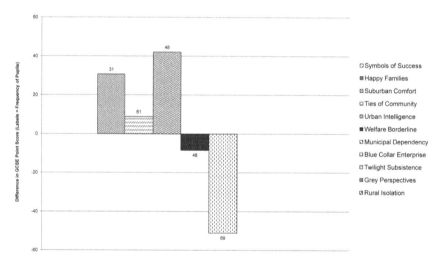

Figure 8.5   Average neighbourhood level GCSE point scores within Whalley Range High School compared with the school average GCSE score

Like Parrs Wood High School, it is those students from more affluent neighbourhood Groups within the school who are achieving above the school average (Figure 8.5). When benchmarked against the national neighbourhood attainment profile, school attainment is above the national average across all neighbourhood Types, except for "Suburban Comfort" (see Figure 8.6).

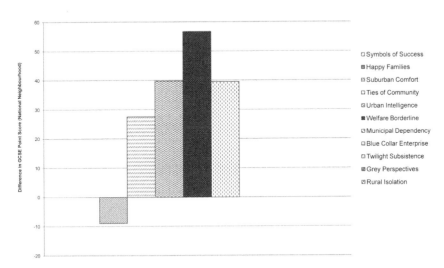

Figure 8.6   Average neighbourhood level GCSE point scores within Whalley Range High School compared with the national neighbourhood average GCSE scores

These patterns of pre-16 attainment affect the pool of students available to study at post-16, and although students do not necessarily directly transfer from with the school, the profiles of both Whalley Range High School and Parrs Wood School at KS5 are nevertheless similar to their KS4 profiles (see Figure 8.7 and Figure 8.8 respectively).

The Whalley Range High School is the lowest performing at KS5, and when compared to the highest performing KS5 school, the King David High School (see Figure 8.9), it is clear that there are considerable differences between the types of areas in which students live. Pupils in the highest performing school almost all live in those affluent neighbourhoods categorised as "Symbols of Success". The lowest performing school has a more mixed profile of less affluent neighbourhoods, with the majority of students living in neighbourhoods classified as "Welfare Borderline".

Parrs Wood High School has average performance at post-16 and the geodemographic profile of students studying there is predominantly overrepresented in "Urban Intelligence" and "Municipal Dependency" neighbourhood Groups.

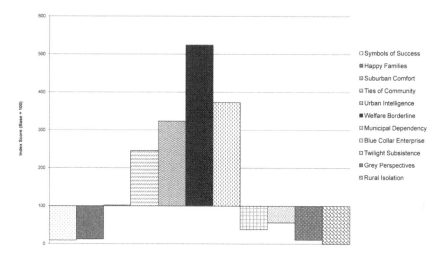

Figure 8.7   Mosaic Profile of the Whalley Range School – KS5

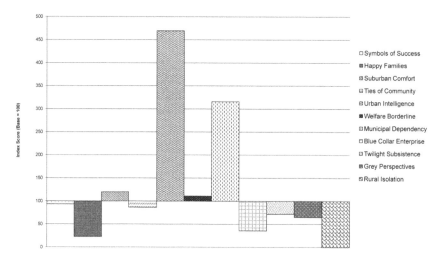

Figure 8.8   Mosaic Profile of Parrs Wood High School – KS5

Although both of these groups are present in the Whalley Range School, there are only average numbers of students from "Welfare Borderline" areas. One explanation for this could be that Parrs Wood High School[1] expects incoming students to possess five A*–C grades at GCSE plus a B grade in the subjects they

---

1   www.parrswood.manchester.sch.uk.

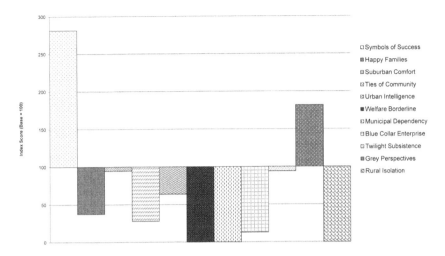

Figure 8.9   Mosaic Profile of King David High School – KS5

wish to study at A-Level. However, Whalley Range 11–18 School[2] has lower entry requirements of five GCSEs at Grade C or above. Thus, if attainment at KS4 varies between neighbourhood Types, this would have an effect on the overall profile of students admitted if these criteria were adhered to strictly.

It is difficult to explain why differential attainment occurs between neighbourhood Groups within a school from empirical research alone; however, it is probably related to the support structures that exist within and between families outside of school. Thus, some schools in Manchester will do more for those students from low participation and achieving neighbourhoods to raise attainment above the national average, whereas other schools may simply add further value to students from high attaining neighbourhoods. For a Higher Education institution targeting schools for widening participation activities, this type of information may be very useful, as although a school may show a high profile for low students from participation neighbourhoods, these students may be relatively poor in attainment.

Spatial Distribution of Advantage – Overlapping Markets?

The previous section has shown how attainment at KS4 and KS5 is stratified between schools and spatially by neighbourhood Type. However, stakeholders may also be interested to know why these neighbourhood patterns exist, and in order to explore these issues, a method of visualising the geographic distribution was required. Thus, educational geodemographics can exploit the toolkit of GIS/ GIScience which has long been used in retail studies of "market areas" (Birkin

---

2   www.whalleyrange.manchester.sch.uk.

et al., 2002). Although an average score of the distance that students travel to study A-Levels is a useful measure of the extent of a school/college attendance area, this neither demonstrates where geographically the students are supplied from, nor measures the competition between school areas of acceptance. Thus, in order to examine the spatial patterns of school recruitment, a series of maps have been created using a kernel density estimation technique (De Smith et al., 2007). This technique measures the density of students within a kernel bandwidth (e.g. 1000m) and outputs a raster grid at a pre-specified resolution (e.g. 250m) containing the average density of students. Additionally, using the implementation of this technique in Hawths Tools,[3] an extension for ArcGIS,[4] it is possible to extract a vector boundary delineating different volume thresholds (the per-cent volume contour, PVC). Such features can be used to identify the area which likely contains a given percentage of the students attending a given school or college. Gibin et al. (2007) illustrate this as shown in Figure 8.10, using a density surface for two points. The horizontal plane through the two kernels centred upon the two points represents the 50th percentile, and is the point at which the PVC area and shape are calculated. The PVC is represented in the left hand diagram using a white line.

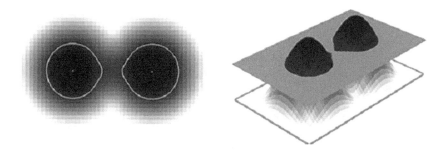

Figure 8.10   50% Kernel density volume contour for a two point dataset
Source: Gibin et al., 2007.

In Manchester, there are 14 schools or colleges which appear on the national pupil database for attainment recording purposes (See Table 8.1). Of these institutions ,three are state schools, seven are independent schools or colleges and four are larger FE colleges. Postcode level data on the independent schools were unobtainable for this study and thus such schools are excluded from the analysis. Furthermore,

---

3   Hawths Tools can be downloaded for free from: www.spatialecology.com/htools.

4   Arc GIS is a software package available from ESRI: www.esri.com/software/arcgis.

Table 8.1 Performance data for Schools and Colleges in Manchester at KS5

| | Number of students aged 16–18 | General and Vocational A/AS or Equivalent Achievement | | | Av pt Score/Entry | | Type | Distance |
		Number at end of A/AS or equivalent study	Average point score per student	Average point score per examination entry	LA Difference	England Difference		
LA Average			658.5	209.6				
England Average			721.5	206.2				
Abbey College	141	39	738	214	4.4	7.8	Ind.	–
Chethams School of Music	125	74	712.3	244	34.4	37.8	Ind.	–
City College Manchester	2939	290	644.9	210.8	1.2	4.6	FE	5.66
The King David High School	225	111	783.1	229.4	19.8	23.2	State	1.54
Loreto College	1592	544	718.2	211.8	2.2	5.6	FE	3.23
Manchester College of Arts and Tech.	3977	332	488	209.6	0	3.4	FE	3.67
The Manchester Grammar School	369	189	1006.6	256.9	47.3	50.7	Ind.	–
Manchester High School for Girls	200	94	1045	252.2	42.6	46	Ind.	–
Parrs Wood High School	411	184	792.2	204.7	-4.9	-1.5	State	1.57
St Bede's College	264	136	941.9	221.4	11.8	15.2	Ind.	–
Whalley Range 11–18 High School	262	76	554.4	182.4	-27.2	-23.8	State	1.56
William Hulme's Grammar School	94	42	872.3	212.4	2.8	6.2	Ind.	–
Withington Girls' School	166	82	1181.7	261.2	51.6	55	Ind.	–
Xaverian College	1326	462	653.1	207.7	-1.9	1.5	FE	2.29

from the LSC data, it was impossible to extract those students who were studying A-Levels alone, so instead the age band 16–19 was used, which bounds the usual age range within which these students would typically be studying.

Using both PLASC and ILR data, the postcode of student homes can be geocoded and mapped. Figure 8.11 shows the location of the schools/colleges considered in the proceeding analysis and the areas from which their students are drawn.

The bands on this map relate to a PVC of 50%; that is, they are estimated to contain 50% of the students attending the attributed school or college. The advantage of using PVC over comparable techniques, such as convex hulls, is that this technique has a lower chance of disclosing the identities of individual students as the distributions are smoothed. Furthermore, isolated pockets of areas which supply students at a high or low propensity can be identified using PVC, whereas convex hulls would simply bound all data within a given frequency of students (e.g. 50%). These distributions show that the FE colleges all seem to be recruiting from the South of Manchester, outside of the local authority, whereas the state schools have more local areas of attendance which are geographically concentrated. This is also reflected in the median distance travelled to attend school or college shown Table 8.1. Figure 8.12 shows a local profile for the King David High School with the additional details of the kernel density estimate raster. This raster varies from dark to light colouring depending on the density of students present within the 250 metre cells. It is possible to identify two peak areas in which most students live.

Figure 8.13 shows the location of students attending Whalley Range High School. The area of attendance for this school is more spread out, with a number of smaller satellite clusters along arterial roads. A similar pattern is exhibited by those students attending Parrs Wood High School (see Figure 8.14). The King David High School is the highest achieving state sector school in Manchester at KS5 (see Table 8.1). The school is also very high achieving for the pre-16 age group with 98% of 15 year olds achieving five or more A*–C grades at GCSE, which would make it very popular for parents choosing a secondary school. The geographical selection criteria used to select students at KS4 suggest an explanation for the reasonably constrained area of attendance in the King David High School at KS5, as many students will transfer into the sixth form after GCSE, and will likely live in the area that is close to the school. Conversely, both the Whalley Range High School and the Parrs Wood High School perform below the local authority and England averages for KS5 (see Table 8.1). The performance for KS4 is also low with the Whalley Range High School achieving 38% 5+ A*–C grades at GCSE and Parrs Wood High School 51%. This is likely to affect the desirability of the school and therefore it is unsurprising that the area of attendance will extend across a larger area.

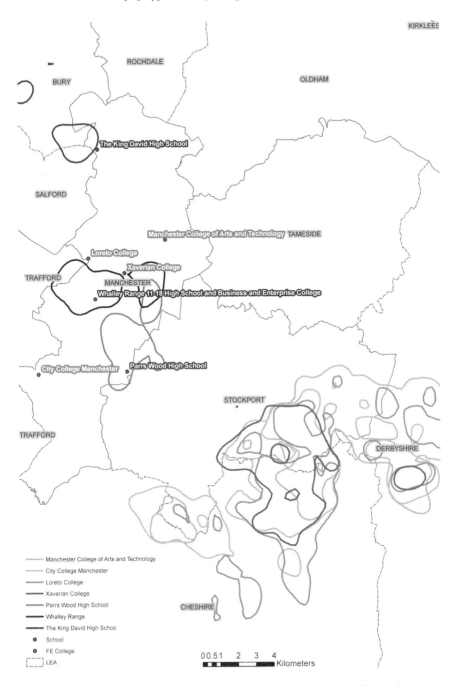

Figure 8.11   Schools/colleges in Manchester and their areas of attendance – KS5

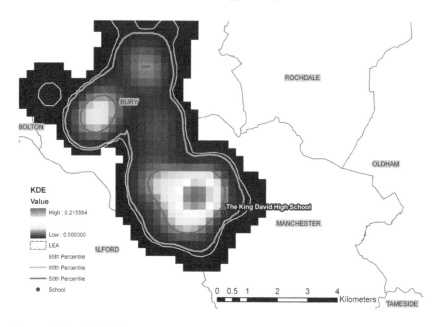

Figure 8.12   The King David High School area of attendance – KS5

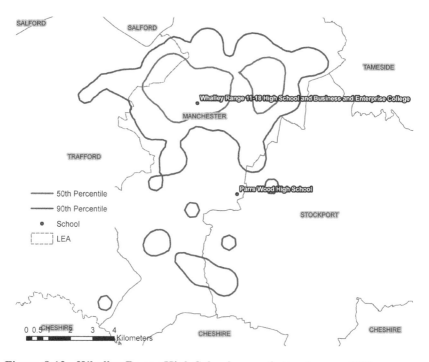

Figure 8.13   Whalley Range High School area of attendance – KS5

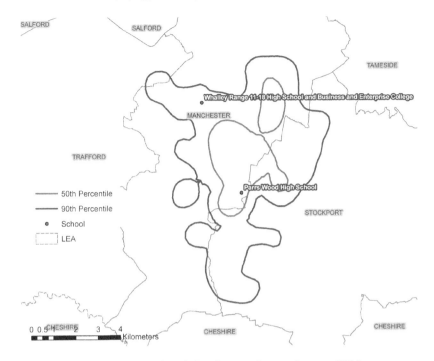

Figure 8.14   Parrs Wood High School area of attendance – KS5

Benchmarking and Investigating Performance for Higher Education Stakeholders

The case study of Manchester highlights how Higher Education institutions may wish to account for neighbourhood level differential attainment in offer and decision making. This is illustrated here by comparing school attainment at KS4 by neighbourhood Group against a national benchmark. Schools can be ranked in terms of their performance at raising student attainment from within these Groups as a value added measure. This could potentially be a useful exercise for a university looking towards conducting outreach activities with pre-16 year students designed to raise aspirations of students who are less likely to participate in Higher Education. The analysis which follows compares the average GCSE point score within a school for those students classified as living within a "Welfare Borderline" neighbourhood to the average score achieved by all students within England for this Group. "Welfare Borderline" was chosen as the target neighbourhood Group; however, this could be adapted depending on the application/aim of the analysis to include different Groups or Types. Those schools which had less than 20 total pupils, and less than 10 students from within "Welfare Borderline" neighbourhood areas, were filtered from the dataset to prevent possible outlier results based on small counts. The remaining schools were ranked by the students from "Welfare

Borderline" performance gain over the national average, and the top three schools are identified in Table 8.2. The purpose of examining these schools individually is to discover those underlying properties which raise average GCSE scores, with the aim of creating a framework for raising overall attainment in England.

Table 8.2   Top three schools by performance gain for students from "Welfare Borderline" neighbourhoods

Rank	School	Point Gain
1	211 Sir John Cass Foundation and Redcoat Church of England Secondary School	233.8
2	208 London Nautical School	224.7
3	316 Plashet School	220.8

The Sir John Cass Foundation and Redcoat Church of England Secondary School has the geodemographic profile shown in Figure 8.15. The school has a very large proportion of students from "Welfare Borderline" neighbourhoods, and as such has found curriculum strategies to raise attainment from this typically low attainment neighbourhood Group. The school has received a large amount of praise in the press surrounding the improvement in performance raising their A*–C grades at GCSE from 22% to 79% (Ward and Smithers, 2004; Eason, 2004). The key to this success is hinted at in a quotation from the head teacher which appeared in the press:

> We used one of our major strengths – because we have a large proportion of students for whom English is their second language. (Eason, 2004)

The school is a specialist language college in a very multicultural area. Instead of just teaching those European languages typically found in English state schools, at this school, pupils can take GCSEs in 10 languages including Bengali, Urdu, Turkish and Russian (Ward and Smithers, 2004). This is an example of a school where they have adapted their curriculum offerings to raise attainment (as measured by performance indicators).

The second highest performing school was the London Nautical School, which has a more affluent geodemographic profile (see Figure 8.16) with a greater proportion of students coming from "Urban Intelligence" neighbourhoods. This school is the highest performing non religious state school in the London Borough of Lambeth. The school is comprehensive in that any student can apply; however, it operates a banded intake, where applicants are divided into groups based on an entry exam. The school is over subscribed across all of these groups; that is, more children take the test than there are places. Thus, within each of the three bands a series of criteria are applied which include:

- A test on nautical interest
- National or County level sporting achievement
- Sibling within the school
- Distance to the school (only where school is not the parents' first choice)

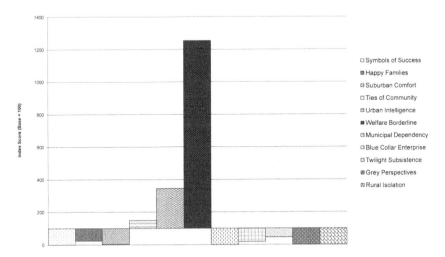

Figure 8.15   Sir John Cass Foundation and Redcoat Church of England Secondary School geodemographic profile – KS4

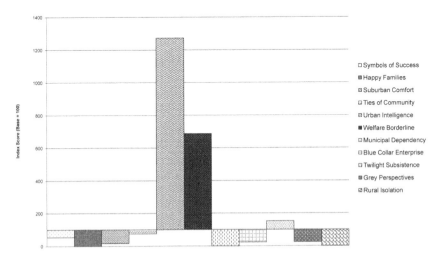

Figure 8.16   London Nautical School geodemographic profile – KS4

Because this school is over subscribed by those who apply to the school as their first choice, distance will only be used in a very small minority of circumstances. Thus, parents who wish for their children to attend this school require a degree of motivation to prepare for the series of tests that are required to gain entry. A successful application is not determined by the Standard Assessment Tests (SATs) aptitude banding, as students are admitted from all bands in equal proportions. Thus the priority for parents wishing for a successful application to this school lies on the "tie breaker" nautical test, be this through genuine or acquired knowledge and interest in nautical activities. This school does not outperform those independent schools within the borough or surrounding areas, although it does provide a good alternative for those families who cannot afford these options or who cannot move within the locality of those high performing state schools with more stringent geographic selection criteria. This explains the under-representation of those students from these more affluent areas that may have chosen to be schooled privately. Thus, the higher performance for students from "Welfare Borderline" neighbourhoods within this school may partially be attributable to the cultural capital related to the support structures that exists within the family home, i.e. the same support structure which led to them gaining a place through a rather complicated and unusual application system.

The final school which will be discussed in this section is Plashet School, which is a girls school within the London borough of Newham. Selection for this school is based on geographic proximity to the student home. This school is the second highest performing state school within the Borough and has the geodemographic profile shown in Figure 8.17. There is a large overrepresentation of students from the neighbourhood Group "Ties of Community", which are areas typically characterised by people with low educational levels and living in close knit communities. A shared similarity between this Group and "Welfare Borderline" is that they both feature an overrepresentation of ethnic minorities in their demographic. The Plashet School website[5] states that "the girls are predominantly of Asian origin", and if the ethnicity of the students are examined from the PLASC data, this is also shown to be true with 37.5% Pakistani, 24.9% Indian and 25.7% Bangladeshi in origin.

There is clearly an ethnic dimension to the raised achievement within this school, and if the England attainment for these groups is averaged by ethnic group (see Table 8.3) one can see that all three of the Asian ethnic groups attain higher than the England national average, which is 360 points. This pattern is replicated within Plashnet school, with all groups also higher than the national average and within their corresponding English ethnic group averages. This higher performance in the "Welfare Borderline" neighbourhood Group seems to be related to the performance of the composition of ethnic groups within the school. Gender differences have not been explored in a national context. However, as Conolly (2007:14) discussed, "gender does tend to exert an influence on GCSE attainment such that boys in general tend to achieve less than girls", although "these differences are relatively small and tend to be overshadowed by the effects of social class and ethnicity".

---

5   www.plashet.newham.sch.uk/staticpage.aspx?prdID=1.

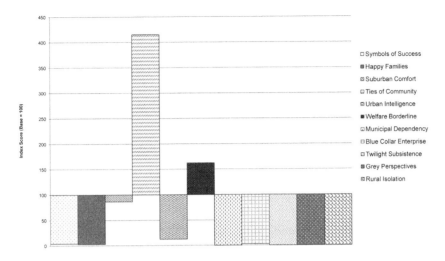

Figure 8.17   Plashet School geodemographic profile – KS4

Table 8.3   Average attainment by ethnic group in England and Plashet School

Ethnic Group	England Average Score	School Average Score	School Average Differences	National Average Differences
Bangladeshi	365	516	+ 151.597	+ 156
Indian	422	545	+ 122.713	+ 185
Pakistani	347	480	+ 133.733	+ 120

Profiling and School Selection Policy

In England, the 1988 Education Reform Act (OPSI, 1988) created a system of "open enrolment" for secondary schools. The underling principle was to "improve schools and provide high quality education for all" (Bauch, 1998:309) through increasing choice available to educational actors (e.g. parents) and introducing funding partially linked to student numbers (Tooley, 1997). Thenceforth, secondary schools no longer operated catchment areas, and all parents were given autonomy to send their children to any school they chose. However, although this model was designed to increase parental choice and equality, the reality was that the majority of parents chose a high performing school, thus resulting in massive over subscription. Although some state schools are selective by attainment, the majority are not, yet operate other selection criteria which largely determine who can gain entry. Common types of selection may be based on religious observance, or in the case of specialist colleges, an aptitude for a particular subject such as Languages,

Sports or Technology. The most common selection strategy used by schools, after religion and subject aptitude, is to select students based on their geographical proximity to the school, as measured by straight line distance from the pupil's home. A limited supply of property surrounding the highest performing schools and an increased demand caused by the geographical selection criteria often pushes up selling prices of the local housing market (see Black, 1999; Gibbons and Machin, 2003; Leech and Campos, 2003). This has the effect of locking out lower socio-economic groups and is an example of what Ball (2003) describes as "the way in which the middle class maintain and improve their social advantages in and through education".

There has been much media attention over issues of KS4 selection and an announcement, that under David Cameron's leadership of the Conservative Party, there will be no extension of selection in state schools except in special cases where demographic change triggers the need to add incrementally to the current stock of academically selective schools. Under this selective system, entry criteria are based on attainment in an exam taken at aged 11 (Willets, 2007). Although a Grammar school indicator does not feature on the PLASC schools database, those schools which operate selective admissions criteria can be identified. The Mosaic geodemographic profile of these schools is shown in Figure 8.18.

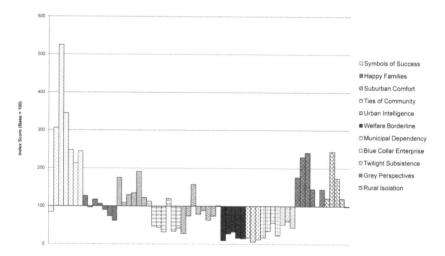

Figure 8.18   2006 KS4 Profile of selective schools in England

This profile has high overrepresentation of those groups which represent the most affluent areas and supports the Conservative Party claim that Grammar schools are "no longer the vehicles for progress for bright children from poor backgrounds" (Willets, 2007). It should however be noted that Grammar Schools account for only around 5% of state school admissions.

The policy disengagement with selection begs questions as to what is a meritocratic system of admissions? Much debate surrounds the various policies which could replace the current system. A report by the Institute for Public Policy Research (IPPR) argued:

> All local authorities should move towards a system of area-wide banding, where the objective of achieving a mixed ability intake of pupils at every school would sit alongside other factors such as parental preference and the distance from home to school. Tough and Brooks. (2007:4)

Such a system would be designed to maximise:

- High levels of pupil attainment
- Fairness for all pupils and parents
- Community cohesion, socialisation and citizenship
- Public trust and satisfaction

A completely random system which allocated places based on a *lottery* should be considered the only truly *comprehensive* system as no bias can be attained by any particular group. However the extent to which this system would be fully accepted and utilised by all members of society is debatable. A British Council funded research project conducted by Jarvis and Alvanides (2007) found that "lotteries of over-subscribed school places would produce the worst of both worlds – greater educational polarisation and longer, more environmentally damaging car journeys to distant schools by middle class parents" because of the unequal ability of parents from different social groups to travel to schools far away from their homes. In Brighton and Hove, a lottery system has been introduced to randomly allocate places to high performing state schools (BBC, 2007). However, the lottery operates within newly designed school catchments created as aggregations of postcode sectors. Students are given priority of access to the school within their local catchment area, and where aggregate demand exceeds school supply of places, the students are allocated to places randomly. This does seem an odd choice of system, as there is mixed frequency of schools within the catchment areas, and as such some students' choice will be restricted to a single school. Furthermore, property premiums which accrue on houses near the most popular schools will still exist because of the catchment criteria, although perhaps not to the same degree in the immediate locality of the most popular schools, because of the abandonment of straight line distance admission criteria.

A School Selection Case Study – The London Borough of Lewisham

The model suggested by the Institute of Public Policy Research (IPPR) is already implemented in a number of Local Authority areas, including the London borough of Lewisham. Haberdashers' Aske's Hatcham College is the most oversubscribed secondary school in Lewisham, and indeed England, with around 2500 applications for 200 places.[6] In this London borough, students progressing into secondary school are allocated into 9 attainment bands (1 = Lowest, 9 = Highest) based on their responses to a Nelson Non Verbal Reasoning Test[7] (NVRT) which is used as part of this school's selection criteria. The criteria used to allocate places at this school are as follows:[8]

1. Students with Statements of Special Educational Needs where the College has consented to be named in the Statement.
2. Students in public care (looked after children).
3. 10% of students are admitted on the basis of aptitude in Music, using a specified assessment process.
4. Students for whom it is essential to be admitted to the College because of special circumstances to do with significant medical or social needs evidenced by written medical advice.
5. Students who, on the date of admission, have a sibling (i.e. a natural brother or sister, or a half brother or sister, or a legally adopted brother or sister or half-brother or sister; who will be living with them at the same address at the date of their entry to the College) on the roll of the Haberdashers' Aske's Hatcham College.
6. Of the remaining places:
   A. 50% are offered to students living within three miles (4.8Kms) and south of the Thames, on the basis of proximity; i.e. students who live the nearest radial distance to the College on the close of the admission application date.
   B. The remaining 50% are offered to students living within three miles (4.8Kms) and south of the Thames, on the basis of an independently operated random allocation.

The random allocation in criterion 6B occurs within the 9 attainment bands based on the scores of the NVRT. Because attainment varies by neighbourhood Groups, this banding also has the effect of mixing up the geodemographic profile of the students who attend the school (Figure 8.19). Haberdashers' Aske's Hatcham College achieved 91% A*–C grades, however, it is also marked by a lower proportion of those geodemographic groups who normally dominate the best schools within an area ("Symbols of Success"). This is partially due to the location of the school ("Urban Intelligence" is the dominant neighbourhood Groups around the school), but also could be attributable to differential mixes of social and human capital between these neighbourhood groups, and the resulting decisions they

---

6 Source: www.wikipedia.org/wiki/Haberdashers'_Aske's_Hatcham_College.
7 The Nelson Non Verbal Reasoning Tests are available from: www.nfer-nelson.co.uk.
8 Source: schools admissions criteria: www.haaf.org.uk/hpropectus-admissions/hprospectus/hadmissionscriteria.htm.

make about choice of school. This links to the those discussions from Chapter 3 and demonstrates how "Urban Intelligence" neighbourhoods will likely have the inclination to choose the best school, however not necessarily the financial means of pursuing their first choice (which may be an independent school in London), thus leaving their only option to choose a state school which will, they believe, provide their child the best future opportunities. This latter hypothesis was discussed earlier in the context of the London Nautical School which has a similar admissions policy, and indeed the geodemographic profiles of the two schools are similar (see Figure 8.19).

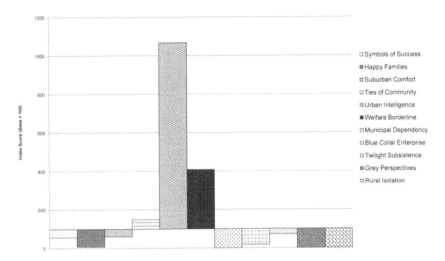

Figure 8.19   Mosaic profile of Haberdashers' Aske's Hatcham College – KS4

The national adoption of randomised selection criteria requires further analysis and modelling to ensure that attainment and quality within the sector do not fall. Further investigation is needed into which are the most appropriate methods of measuring students' latent ability, how many bands should this ability be divided into and the frequency of students that should be placed within them. Finally, from a policy discourse perspective, "piecemeal social engineering" (Popper, 1971) which restricts a school's ability to create their own selection criteria should be examined against other free market models of school place allocation, such as voucher schemes.

Conclusion

This chapter develops a series of case studies relevant to a range of stakeholders in Higher Education. These directly relate to a series of analysis that these groups of people may wish to perform including

- Exploring the effects of different school selection policy
- Investigating operational issues such as selecting the most appropriate schools to target with widening access initiatives
- Performing geographical analysis of school market areas
- Creating school performance benchmarks to target resources to raise attainment in low participation neighbourhoods

Furthermore, GIS/GIScience functionality has been shown to be relevant to extending educational geodemographics to visualise geographic market areas of schools.

Chapter 9
Conclusion: The Geography of Access to Higher Education

This book has examined the multiple dimensions of social, spatial and temporal processes which shape access to Higher Education. It has concluded that there are distinct social inequalities of access, and that these have distinctive geographic manifestations that are relevant to improving access to the sector. Full and contextualised understanding of the full range of social, economic and spatial concepts is acutely important to stakeholders (potential students, universities, schools) in Higher Education. Recent changes in the policy and funding of Higher Education, in the UK and elsewhere, have presented institutions with the following challenges:

- To demonstrate how sector and institutional data might enable universities and colleges to position themselves within a changing policy framework.
- To understand how the information that institutions themselves collect may be better utilised to help institutions design their marketing and student support strategies.
- To identify the geographical areas from which institutions should be able to attract students.
- To analyse the nature of the disciplines offered within institutions and the types of students they attract.

The agendas of widening participation, extending access and institutional marketing share common challenges to devise better ways of reaching potential students who are appropriately qualified and motivated to pursue and successfully complete the full range of institutional course offerings. This book has set out some relevant aspects of the changing Higher Education policy-setting arena and presented a systematic framework for widening participation and extending access in an era of variable fees. In particular, this has aimed to illustrate how Higher Education data and publicly available sources might be enriched to enable institutions to move from piecemeal analysis of their intakes, to institution wide strategic and geographically linked market area analysis for existing and envisaged subject and course offerings.

Historically, most Higher Education systems have served privileged minorities, although more recent Higher Education policies have, to varying extents, transformed the sector into a system which provides for a larger and more socially representative cross section of society. Through the introduction of student fees,

Higher Education is rapidly developing some of the characteristics of a traditional commercial market, thus increasing demand for information by both consumer and producer stakeholders. Despite rapid growth in the absolute numbers of students attending Higher Education, it is apparent that there remain access inequalities between students from different socio-economic backgrounds. Monitoring of the Higher Education sector involves numerous organising bodies, each with their own remit, data and collection mechanisms: the result, in the UK at least, is an educational data economy characterised by overlaps, lack of coordination and missing data. There is little in the way of the kinds of centralised and customer orientated services which are essential and relevant for making decisions in Higher Education. Thus, an important message of this book is that there is an urgent need to systematically integrate and bring relevant data together for the benefit of Higher Education stakeholders.

The data analysis presented in this book has demonstrated that access to Higher Education is both spatially and socially complex. The distance that applicants travel to undertake courses of Higher Education in the UK is related to the specialism of a course, the type of institution to which a candidate applies, and the tariff score required of the applicants. Higher Education qualifying attainment scores differ between neighbourhood Types, and higher attainment is predominantly observed in students from more affluent neighbourhoods. The national indicators and benchmarks used to assess performance, and allocate funding to institutions for widening participation, are shown to be of variable quality in terms of identifying the neighbourhoods least likely to participate in Higher Education. The complexity of the education sector has been illuminated through spatial analysis, and highlights a critical need for better information and decision support tools.

The addition of Higher Education data into a geodemographic classification with open methodology is a positive step beyond the use of generic and re-labelled commercial classifications, and presents an improvement of use of classifications for purposes beyond those for which they were not originally designed. This provides a challenge to the implied assumption that individuals use public services, such as Higher Education, in ways that are directly analogous to consumption of private goods. These analyses have demonstrated the first bespoke geodemographic classification created specifically for Higher Education applications using public sector data. A further technical and computational issue investigated was the need to understand how the output from a k-means clustering algorithm might be made more robust through re-running models multiple times to extract an optimised solution. Evaluation of geodemographic classifications has demonstrated some of the unique and shared properties between the commercial classification Mosaic and OAC. Through these comparisons, it has been possible to see which Mosaic Groups are formed from data related to income, and the relative information loss that one might expect from the switch between a classification that categorises neighbourhoods at OA rather than unit postcode scale. Using a series of evaluation methods, the performance of classifications were compared, in order to effectively partition both course and institution access rates and to predict levels of aggregate participation. Although the commercial classification demonstrates marginally better performance, possible issues with the way in which the typology stratifies neighbourhoods have been highlighted by this analysis.

The final part of the book examined whether inequalities are getting better or worse and how these could be addressed through engaging with the stakeholders in Higher Education. These analyses extend from previous analysis to examine longitudinal trends related to participation rates and educational histories of potential applicants. Discussion relates a need to theorise the links between supply and demand in order to suggest routes towards more efficient and effective distribution of the life chances of students, in order better to accommodate their likely needs, preferences and the uniqueness of places within generalised representations of the system. It was shown how the Higher Education sector has variable growth rates between courses within institutions, and that these changes are not evenly distributed across societal groups. Furthermore, it was illustrated how socio-economic status, which is relied upon for much sociological and educational research in the area of access to Higher Education, is slowly being eroded as a useful indicator through increasing rates of incomplete data. Using area classification, access to Higher Education from low participation neighbourhoods was shown to have improved marginally over the period 2001–2004, although strong inequality in participation rates between those neighbourhoods most and least likely to attend Higher Education still remain. The final investigation presented an analysis of case studies relevant to a range of stakeholders in Higher Education.

The thematic threads linking these concluding comments develop a picture of a complex and dynamic sector with intense and growing informational demands, yet which currently lack any centrally provided service to meet their requirements. Many Higher Education institutions are currently making uninformed choices related to widening participation, which is of acute policy concern, and there still remains a requirement for much improved decision support tools. There is a highlighted need to measure and monitor changes in this fast developing field and even more so than in the traditional applications domains of geodemographics. This research begins to address these concerns through the creation of bespoke measures of classifying participation rates using disparate educational and contextual databases, with demonstrated application of how stakeholders can adopt more appropriate strategies to target and accelerate improvement in access inequalities.

References

AHERG (2004) Fair Admissions to Higher Education: Recommendations for Good Practice. London: Admissions to Higher Education Review Group.

Ainley, P. (1994) *Degrees of Difference, Higher Education in the 1990s*. London: Lawrence and Wishart.

Allen, D. (2005) A Delightful Sport with Peculiar Claims: Otter Hunting in Britain, 1880–1939. 31st Royal Geographical Society Annual International Conference. 31 August–2 September 2005. London: RGS.

Allison, R. (2002) Universities to be set Quotas for Poorer Students [online]. Available from: www.education.guardian.co.uk/higher/news/story/0,9830,857203,00.html [Accessed 10 December 2002]. London: The Guardian.

Anderson, R.C. (1977) The Notions of Schemata and the Educational Enterprise: General Discussion of the Conference. In R.C. Anderson, R.J. Spiro, W.E. Montague (eds) *Schooling and the Acquisition of Knowledge*. New Jersey: John Wiley and Sons.

Anderson, R.D. (1992) *Universities and Elites in Britain since 1800*. London: Macmillan.

Archer, L., Hutchings, M., Ross, A. (2003) *Higher Education and Social Class: Issues of Exclusion and Inclusion*. London: Routledge.

Ashby, D.I., Longley, P.A. (2005) Geocomputation, Geodemographics and Resource Allocation for Local Policing. *Transactions in GIS*. 9 (1), 53–72.

Ashton, M. (2005) The Family Car: The Geographies of In-car Space. Royal Geographical Society Annual International Conference. 31 August–2 September 2005. London: RGS.

Aston, L., Bekhradnia, B. (2003) *Demand for Graduates: A Review of the Economic Evidence*. London: Higher Education Policy Institute.

Atkinson, R., Bridge, G. (2005) *Gentrification in a Global Context: The New Urban Colonialism*. London: Routledge.

Aveyard, P., Manaseki, S., Chambers, J. (2000) Does the Multidimensional Nature of Super Profiles Help District Health Authorities Understand the way Social Capital Affects Health? *Journal Of Public Health Medicine*. 22, 317–323.

Ball, S.J., Bowe, R., Gewirtz, S. (1996) School Choice, Social Class and Distinction: the Realization of Social Advantage in Education. *Journal of Education Policy*. 11(1), 89–112.

Ball, S. (2003) *Class Strategies and the Educational Market*. London: Routledge Falmer.

Barker (2006) National Pupil Database: Current Content and Structure [online] Available from: www.bris.ac.uk/depts/CMPO/PLUG/userguide/anna.ppt [Accessed 10 June 2006].

Bassett, K., Short, J. (1980) *Housing and Residential Structure: Alternative Approaches*. London: Routledge.

Bauch, P. A. (2000) Do School Markets Serve the Public Interest? More Lessons from England. *Educational Administration Monthly*. 36, 309.

BBC (2002) Fee Paying Pupils get the Worst Degrees [online]. Available from: www.news.bbc.co.uk/1/hi/education/2552523.stm [Accessed 28 November 2003] London: BBC.

BBC (2004) Four for the Price of Three [online]. Available from: www.news. bbc.co.uk/go/pr/fr/-/1/hi/education/3436961.stm [Accessed 28 April 2004]. London: BBC.

BBC (2006) £2m to Halt Decline in Geography [online]. Available from: www. news.bbc.co.uk/2/hi/uk_news/education/4860226.stm [Accessed 16 June 2007]. London: BBC.

BBC (2007) Schools to Give Places by Lottery [online]. Available from: www.news. bbc.co.uk/1/hi/education/6403017.stm [Accessed 1 June 2007]. London: BBC.

Bell, D. (2004) The Value and Importance of Geography. *Primary Geographer* 56, 4–5.

Birkin, M. (1995) Customer Targeting, Geodemographics and Lifestyle Approaches in P. Longley, G.P. Clarke (eds) *GIS for Business and Service Planning*. Cambridge: Geoinformation.

Birkin, M., Clarke, G.P., Clarke, M. (2002) *Retail Geography and Intelligent Network Planning*. London: Wiley.

Black, S.E. (1999) Do Better Schools Matter? Parental Valuation of Elementary Education. *Quarterly Journal of Economics*. 114, 78–599.

Blair, T. (2000) *Ambitions for Britain*. Labour's Manifesto 2001. London: Labour.

Blair, T. (2003) Big Conversation [online]. Available from: www.bigconversation. org.uk/index.php?id=684 [Accessed 1 January 2004]. London: Labour Party.

Blanden, J., Gregg, P., Machin, S. (2005) *Intergenerational Mobility in Europe and North America*. London: The Sutton Trust.

Blasko, Z. (2002) *Access to What: Analysis of Factors Determining Graduate Employability*. November 2002. London: Centre for Higher Education Research and Information.

Blunt, W. (2001) *Linnaeus: The Complete Naturalist*. London: Frances Lincoln.

Boliver, V. (2005) Classification of Higher Education Institutions. Email message to: Vickie Boliver. 18 January 2005. Personal Communication.

Borges, J.L. (1975) *Other Inquisitions: 1937–1952*. Texas: University of Texas Press.

Bourdieu, P. (1986) The Forms of Capital. In J.G. Robinson (eds) *Handbook of Theory and Research for the Sociology of Education*. New York: Greenwood Press.

Bourdieu, P. (1993) *Sociology in Question*. London: Sage.

Bourdieu, P., Passeron, J.C. (1977) *Reproduction in Education, Society and Culture*. London: Sage.

Bourdieu, P. and Wacquant, L. (1992) *An Invitation to Reflexive Sociology*. Cambridge: Polity.

Bowers, K., Hirschfield, A. (1999) Exploring Links between Crime and Disadvantage in North-West England: An Analysis Using Geographical Information Systems. *International Journal of Geographical Information Systems*. 13(2), 159–184.

Bowker, G.C., Leigh Star, S. (1999) *Sorting Things Out: Classification and its Consequences*. Massachusetts: MIT Press.

Bracken, I. (1981) *Urban Planning Methods: Research and Policy Analysis*. New York: Methuen.

Brown, R., Piatt, W. (2001) Funding Widening Participation in Higher Education: A Discussion Paper. London: IPPR/CIHE.

Burgess, E.W. (1925) The Growth of the City. In R.E. Park, E.W. Burgess, R.D. McKenzie (eds) *The City*. 47–62. Chicago: University of Chicago Press.

Cadwallader, M.T. (1979) Problems in Cognitive Distance and their Implications to Cognitive Mapping. *Environment and Behaviour*. 11, 559–76.

Carr, F. (1998) The Rise and Fall of The Polytechnics: Explaining Changes in British Higher Education Policy Making. *Policy and Politics*. 26(3), 273–290.

Carr-Hill, R.A., Sheldon, T. (1991) Designing a Deprivation Payment for General Practitioners: the UPA Wonderland. *British Medical Journal*. 302, 393–396.

Carstairs, V., Morris, R. (1991) Deprivation, Mortality and Resource Allocation. *Community Medicine*. 11, 364–372.

Clare, J. (2002) Dropout Students Cost £150m. . 18 January 2002. London: *The Daily Telegraph*.

Cobban, A.R. (1999) *English University Life in the Middle Ages*. London: UCL Press.

Coleman, J. (1988) Social Capital in the Creation of Human Capital. *The American Journal of Sociology*. 94, S96–S120.

Collins, N. (2003) More Cash for Universities? Simple: Shut the Worst Ones [online]. Available from: www.telegraph.co.uk/opinion/main.jhtml?xml=%2F opinion%2F2003%2F01%2F27%2Fdo2702.xml [Accessed 27 January 2003]. London: *The Daily Telegraph*.

Comptroller and Auditor General (2008) *Widening Participation in Higher Education*. National Audit Office, London.

Conolly, P. (2007) The Effects of Social Class and Ethnicity on Gender Differences in GCSE Attainment: A Secondary Analysis of the Youth Cohort Study of England and Wales 1997–2001. *British Educational Research Journal*. 32(1), 3–21.

Creedy, J. (1994) Financing Higher Education: Public Choice and Social Welfare. *Fiscal Studies*. 15(3), 87–108.

Davis, P.M. (1991) *Cognition and Learning: A Review of the Literature with Reference to Ethnolinguistic Minorities*. Dallas: Summer Institute of Linguistics.

Dearing, R. (1997) *The Dearing Report*. Norwich: HMSO.

Debenham, J. (2001) Understanding Geodemographic Classification: Creating the Building Blocks for an Extension [online]. Available from: www.geog. leeds.ac.uk/wpapers/02-1.pdf [Accessed 1 July 2006]. Leeds: The School of Geography, University of Leeds Working Paper.

Department for Education. (1991) *Higher Education: A New Framework.* London: HMSO.

De-Smith, M., Goodchild, M., Longley, P. (2007) *Geospatial Analysis: A Comprehensive Guide to Principles Techniques and Software Tools.* East Sussex: Winchelsea Press.

DfES (1999) Blunkett Welcomes Blair's Higher Education Ambitions [online]. Available from: www.dfes.gov.uk/pns/DisplayPN.cgi?pn_id=1999_0612 [Accessed 12 June 2006]. London: DfES.

DfES (2003) The Future of Higher Education. CM5735. London: HMSO.

DfES (2004) 14–19 Curriculum and Qualifications Reform: Final Report of the Working Group on 14–19 Reform: Summary for Higher Education Institutions. London: DfES.

Draper, D., Gittoes, M. (2004) Statistical Analysis of Performance Indicators in UK Higher Education. *Journal of the Royal Statistical Society, Series A.* 167, 449.

Eason, G. (2004) The Best and Worst Results [online]. Available from: www. news.bbc.co.uk/1/hi/education/3397921.stm [Accessed 1 June 2007]. London: BBC.

Edwards, R., Franklin, J., Holland, J. (2003) Families and Social Capital: Exploring the Issues. South Bank University: Families and Social Capital ESRC Research Group. Working Paper 1, 2003.

Egerton, M., Halsey, A.H. (1993) Trends by Social Class and Gender in Access to Higher Education in Britain. *Oxford Review of Education.* 19(2), 183–196.

Elliot-Major, L. (2007) Most Leaders Privately Educated [online]. Available from: www.news.bbc.co.uk/2/hi/uk_news/education/6246152.stm [Accessed 1 July 2007]. London: BBC.

Eurydice (2004) Two Decades of Reform in Higher Education in Europe: 1980 Onwards [online] Available from: www.eurydice.org/ressources/eurydice/pdf/ 008DN/008_UN_EN.pdf [Accessed 6 July 2004]. Slough: Eurydice Unit for England, Wales and Northern Ireland.

Everitt, B. (1974) *Cluster Analysis.* London: Heinemann Educational Books.

Experian (2006) A Guide to the Mosaic Classification [online] Available from: www. business-strategies.co.uk/upload/downloads/mosaic%20uk%20brochure.pdf [Accessed 5 July 2007]. London: Experian.

Farr, M. (2002) Extending Participation in Higher Education – An Investigation into Applicant Choice Using Postcode Analysis. Unpublished PhD Thesis. Lancaster University.

Forsyth, A., Furlong, A. (2000) *Socioeconomic Disadvantage and Access to Higher Education.* London: Policy Press.

Fotheringham, S. (1997) Trends in Quantitative Methods I: Stressing the Local. *Progress in Human Geography.* 21(1), 88–96.

Fotheringham, S., Wong, D.W.S. (1991) The Modifiable Areal Unit Problem in Multivariate Statistical Analysis. *Environment and Planning A.* 23(7),1025–1044.

Fotheringham, S.A., Brunsdon, C., Charlton, M. (2000) *Quantitative Geography.* London: Sage.

Friedman, M. (1957) *A Theory of the Consumption Function.* Princeton: National Bureau of Economic Research.

Galindo-Rueda, F., Marcenado, O.G., Vignoles, A. (2004) *The Widening Socio-Economic Gap in UK Higher Education.* London: Centre for the Economics of Education.

Gehrman, E. (2006) Geography Center Launched – Center for Geographical Analysis to Explore "vast intellectual territory" [online]. Available from: www.news.harvard.edu/gazette/2006/05.11/05-geography.html [Accessed 17 June 2007]. Harvard: *Harvard University Gazette.*

Gibbons, S., and Machin, S. (2003) Valuing English Primary Schools. *Journal of Urban Economics.* 53, 197–219.

Gibin, M., Longley, P., Atkinson, P. (2007) Kernel Density Estimation and Percent Volume Contours in General Practice Catchment Area Analysis in Urban Areas. GISRUK 2007, NUI Maynooth 11–13 April 2007.

Gini, C. (1912) Variabilità e Mutabilità. In: E. Pizetti, T. Salvemini (eds). *Memorie di Metodologica Statistica.* Rome: Libreria Eredi Virgilio Veschi (1955).

Goddard, A. (2004) Subjects Slot into Class Divides. 30 April 2004. London: *The Times Higher Education Supplement.*

Goldstein, H., Spiegelhalter, D.J. (1996) League Tables and their Limitations: Statistical Issues in Comparisons of Institutional Performance. *Journal of the Royal Statistical Society. Series A.* 159(3), 385–443.

Gordon, A.D. (1981) *Classification: Methods for the Exploratory Analysis of Multivariate Data.* Norwell: Kluwer Academic Publishers.

Goss, J. (1995) Marketing the New Marketing. The Strategic Discourse of Geodemographic Information Systems. In J. Pickles (eds) *Ground Truth.* 130–170. New York: Guildford Press.

Grimson. J., Dobson, J. (2002) So Which One is Privileged? 6 October 2002. London: *The Sunday Times.*

Guardian (2002) University Admissions Favours the Poor [online]. Available from: www.education.guardian.co.uk/higher/news/story0,9830,802329,00.html [Accessed 11 November 2002]. London: *Guardian.*

Hackett, G. (2003) Cambridge Asks to See A-Level Marks. 17 August 2003. 6. London: *The Sunday Times.*

Hall, J.C. (2001) *Retention and Wastage in FE and HE.* Edinburgh: SCRE.

Harris, R. (1998) Considering (Mis-) Representation in Geodemographics and Lifestyles [online]. Available from: www.geocomputation.org/1998/82/gc_82.htm [accessed 30 October 2003]. Bristol: School of Geographical Sciences.

Harris, R. (2001) The Diversity of Diversity: Is there still a Place for Small Area Classifications? *Area.* 33(3), 329–336.

Harris, R., Sleight, P., Webber, R. (2005) *Geodemographics: GIS and Neighbourhood Targeting*. London: John Wiley and Sons.

Harris, R., Longley, P. (2005) Targeting Clusters of Deprivation within Cities. In J. Stillwell, G. Clarke (eds) *Applied GIS and Spatial Analysis*. London: John Wiley and Sons.

Harris, R., Johnston, R., Burgess (2007) Neighbourhoods, Ethnicity and School Choice: Developing a Statistical Framework for Geodemographic Analysis. *Population Research and Policy Review*.

Harvey, D. (1973) *Social Justice and the City*. London: Arnold.

Harvey, J. (2000) *Urban Land Economics* (5th Edition). Basingstoke: Macmillan.

Hensher, D.A., Johnson L.W. (1981) *Applied Discrete-Choice Modelling*. London: John Wiley and Sons.

HESA (2004) Guide to PI [online]. Cheltenham: HESA. Available from: www.hesa.ac.uk/pi/0405/guide.htm [Accessed 12 June 2006]. Cheltenham: HESA.

HEFCE (2004) Understanding of Market Position Crucial to Future Success. HEFCE Council Briefing, 53. Bristol: HEFCE.

HEFCE (2005) Young Participation in Higher Education [online]. Available from: www.hefce.ac.uk/pubs/hefce/2005/05_03/ [Accessed 8 June 2007]. Bristol: HEFCE.

HESA (2006) HESA – Higher Education Statistics Agency [online]. Available from: www.hesa.ac.uk [Accessed 5 July 2007]. London: Labour Party.

HEFCE (2007a) Review of Performance Indicators [online]. Available online from: www.hefce.ac.uk/pubs/hefce/2007/07_12/07_12.pdf [Accessed 23 June 2007]. Bristol: HEFCE.

HEFCE (2007b) Higher Education Outreach: Targeting Disadvantaged Learners [online]. 14 June 2007. Available online from: www.hefce.ac.uk/pubs/hefce/2007/07_12/07_14.pdf [Accessed 23 June 2007]. Bristol: HEFCE.

HEFCE (2007c) Funding Higher Education in England: How HEFCE allocates its Funds [online]. Available online from: www.hefce.ac.uk/pubs/hefce/2007/07_20/07_20.pdf [Accessed 27 August 2007].

Hodge, M. (2002a) New Evidence Shows Degrees Boost Quality of Life [online]. Available from: www.dfes.gov.uk/pns/DisplayPN.cgi?pn_id=2002_0157 [Accessed 31 July 2002]. London: DfES Press Realease.

Hodge, M. (2002b) Poor not needed to reach 50% target, Hodge admits. London: The Times Higher Education Supplement. Available from: www.thes.co.uk/search/story.aspx?story_id=81920 [accessed 12 June 2006].

Hoyt, H. (1939) *The Structure of Growth of Residential Neighbourhoods in American Cities*. Washington DC: Federal Housing Authority.

Jarman, B. (1983) Identification of Underprivileged Areas. *British Medical Journal*. 286, 1705–1708.

Jarvis, H., Alvanides, S. (2007) The School Choice Project [online]. Newcastle: Newcastle University. Available from: www.staff.ncl.ac.uk/s.alvanides/schoolchoice/ [Accessed 1 June 2007].

Jenkins, A. (1995) *Accountable to None: Tory Nationalization of Britain*. London: Penguin Books Ltd.

Johnson, B. (2007) So I Uncorked a Gaffe...but why was it a Gaffe? [online]. Available from: www.education.guardian.co.uk/higher/comment/story/0,,2048476,00.html [Accessed 5 August 2007]. London: *Guardian*.

Jones, P., Elias, P. (2006) Administrative Data as Research Resources: A Selected Audit (online). Available online from: www.rss.org.uk/pdf/Admin%20Data%20selected%20audit%20report%20v2.pdf [Accessed 6 August 2007].

Kearns, R. (2005) Conceptualising the Place of Pet Therapy. Royal Geographical Society Annual International Conference. 31 August–2 September 2005. London: RGS.

Kelly, A. (1976) Family Background, Subject Specialization and Occupational Recruitment of Scottish University Students: Some Patterns and Trends. *Higher Education*. 5(2), 177–188.

Kendall, M.G. (1966) Discrimination and Classification. In P. Krishnaiah (eds) *Multivariate Analysis*. New York: Academic Press.

Kitchin, R., Blades, M. (2002) *The Cognition of Geographic Space*. London: I.B. Tauis.

Knox, P. (1982) *Urban Social Geography: An Introduction*. 2nd Edition. Harlow: Longman Scientific and Technical.

Kosko, B. (1994) *Fuzzy Thinking*. London: Flamingo.

Krech, D., Crutchfield, R.S., Ballachey, E.L. (1962) *Individual in Society*. London: McGraw-Hill.

Lakoff, G. (1987) *Women, Fire, and Dangerous Things: What Categories Reveal about the Mind*. Chicago: The University of Chicago Press.

Lampl, P. (2007) Most Leaders Privately Educated [online]. Available from: www.news.bbc.co.uk/2/hi/uk_news/education/6246152.stm [Accessed 1 July 2007]. London: BBC.

Leathwood, C., Hutchings, M. (2003) Entry Routes to Higher Education. In L. Archer, M. Hutchings, A. Ross (eds) *Higher Education and Social Class*. London: Routledge Falmer.

Leech, D., Campos, E. (2003) Is Comprehensive Education Really Free?: A Case-Study of the Effects of Secondary School Admissions Policies on House Prices in one Local Area. *Journal of the Royal Statistical Society: Series A*. 166 (1), 135–154.

Leventhal, B. (1995) Evaluation of Geodemographic Classifications. *Journal of Targeting, Measurement and Analysis for Marketing*. 1–11.

Lightfoot, L. (2007) Student Checks are "Social Engineering" [online]. Available from: www.telegraph.co.uk/news/main.jhtml?xml=/news/2007/03/08/nuniv108.xml [Accessed 16 July 2007]. London: *The Daily Telegraph*.

Longley, P.A., Webber, R. (2003) Geodemographic Analysis of Similarity and Proximity: Their Roles in the Understanding of the Geography of Need. In P.A. Longley and M. Batty (eds) *Advanced Spatial Analysis: The CASA Book of GIS*. 233–266. Redlands: ESRI Press.

Longley, P.A., Goodchild, M.F., Maguire, D.J., Rhind, D.W. (2005) *Geographic Information Systems and Science*. Chichester: John Wiley and Sons.

Longley, P.A., Goodchild, M.F. (2007) The Use of Geodemographics to Improve Public Service Delivery. In J. Hartley, C. Skelcher, C. Donaldson, G. Boyne (eds) *Managing Improvement in Public Service Delivery: Progress and Challenges*. Cambridge: Cambridge University Press.

Lorenz, M.O. (1905). Methods of Measuring the Concentration of Wealth. *Publications of the American Statistical Association*. 9: 209–219.

Lösch, A. (1954) *The Economics of Location*. New Haven: Yale University Press.

Macleod, D. (2003) Middlesex University offers £1,000 a Year Scholarship to B-Grade Applicants [online]. Available from: www.education.guardian.co.uk/students/tuitionfees/story/0,12757,1073340,00.html [Accessed 21 October 2003]. London: *Guardian*.

Macleod, D. (2004) HE Records Income Rise and Deficit Fall [online]. Available from: www.education.guardian.co.uk/administration/story/0,9860,1163042,00.html [Accessed 16 July 2007]. London: *Guardian*.

MacQueen, J.B. (1967) *Some Methods for Classification and Analysis of Multivariate Observations*. Proceedings of 5th Berkeley Symposium on Mathematical Statistics and Probability. Berkeley: University of California Press. 1, 281–297.

Mander, K. (2007) Demise of Computer Science Exaggerated [online]. Available from: www.bcs.org/server.php?show=ConWebDoc.10138 [Accessed 15 June 2007]. London: British Computer Society.

Marsh, P., Carlisle, R., Avery, A.J. (2000) How much does Self Reported Health Status, measured by SF-36 vary between Electoral Wards with Different Jarman and Townsend Scores? *The British Journal of General Practice*. 50(457), 630–634.

Martin, D. (1991) *Geographic Information Systems and their Socioeconomic Applications*. London: Routledge.

Massimo, C., Haining, R., Signoretta, P. (2001) Modelling High-Intensity Crime Areas in English Cities. *Urban Studies*. 38(11), 1921–1941.

May, T. (2003) *I am Against a Tax on Learning*. Cherwell. 23 May 2003. 20. Oxford: Cherwell.

Modood, T. (1993) The Number of Ethnic Minority Students in British Higher Education: Some Grounds for Optimism. *Oxford Review of Education*. 19(2), 193–182.

Newby, H. (2005) More Warnings Over Decline in Science Students [online]. Available form: www.education.guardian.co.uk/universitiesincrisis/story/0,,1595862,00.html [Accessed 15 June 2007]. London: *Guardian*.

Openshaw, S. (1984) Ecological Fallacies and the Analysis of Areal Census Data. *Environment and Planning A*. 16(1), 17–31.

OPSI (1988) Education Reform Act 1988 [online] Available from: www.opsi.gov.uk/acts/acts1988/Ukpga_19880040_en_1.htm [Accessed 29 July 2007]. London: OPSI.

Palfreyman, D. (2004) The Economics of Higher Education: Affordability and Access Costing and Accountability [online]. Available from: www.oxcheps. new.ox.ac.uk/MainSite%20pages/Resources/EconHEprot.pdf. OxCHEPS Occasional Paper No.10. Oxford: OxCHEPS.

Parker, S., Burrows, R., Gane, N., Hardey, M., Woods, B., Ellison, N. (2007) *The Spatialisation of Class and the Automatic Production of Space*. Forthcoming – Information, Communication and Society.

Peston, M. (1972) *Public Goods and the Public Sector*. London: Macmillan Press.

Phoenix, D. (2003) Government Policy and Higher Education. *Journal of Biological Education*. 37 (3), 108–109.

Popper, K. (1971) *The Open Society and Its Enemies*. New Jersey: Princeton University Press.

Prospects (2002) Widening Participation in Higher Education [online]. Available from: www.prospects.ac.uk/cms/ShowPage/Home_page/Labour_market_information/Graduate_Market_Trends/Widening_participation_in_higher_education__Spring_02_/p!eLdXgb [Accessed 13 June 2006]. London: Graduate Market Trends.

Putnam, R. (1995) Bowling Alone. *Journal of Democracy*. 6(1), 65–79.

Ramesh, R. (2004) Don't Cut Our Fees. 26 February 2004. London: *The Guardian*.

Reay, D., Miriam, D.E., Ball, S. (2005) *Degrees of Choice: Social Class, Race and Gender in Higher Education*. Staffordshire: Trentham Books.

Reid, I. (1998) *Class in Britain*. Malden: Blackwell.

RGS (2006) Geography in Action 2006–2008 [online]. Available from: www. rgs.org/NR/rdonlyres/1F2C5DE8-A3E4-47A3-B30E-270F5EBFF629/0/GeoginAction_AW.pdf [Accessed 17 June 2007]. London: RGS.

RGS (2007) What is Geography? [online] Available from: www.rgs.org/GeographyToday/What+is+geography.htm [Accessed 13 June 2006]. London: RGS.

Robbins, L.C. (1963) Higher Education: Report of the Committee Appointed by the Prime Minister Under the Chairmanship of Lord Robbins 1961–3. London: HMSO.

Robinson, W.S. (1950) Ecological Correlations and Behaviour of Individuals. *American Sociological Review*. 15, 351–357.

Romesburg, H.C. (1984) *Cluster Analysis for Researchers*. Belmont, CA: Lifetime Learning Publications.

Rosch, E., Mervis, C.B., Gray, W.D., Johnson, D.M., Boyes-Braem, P. (1976) Basic Objects in Natural Categories. *Cognitive Psychology*. 8, 382–439.

Rose, D. (1995) A Report on Phase 1 of the ESRC Review of Social Classifications. Swindon: ESRC.

Rose, D., Pevalin, D.J. (2003) *A Researcher's Guide to the National Statistics Socio-economic Classification*. London: Sage.

Rosenberg, M.J. (1956) Cognitive Structure and Attitudinal Affect. *Journal of Abnormal Social Psychology*. 53, 367–372.

Ross, A. (2003) Higher Education and Social Access: To the Robbins Report. In L. Archer, M. Hutchings, A. Ross (eds) *Higher Education and Social Class.* London: Routledge Falmer.

Ryan, A. (2004) Opinion. 29 October 2004. 13. London: *The Times Higher Education Supplement.*

Sanders, C. (2004) Pair may Undercut Pack with £2,000 Fee. 3 December 2004, 1. London: *The Times Higher Education Supplement.*

Scott, P. (2002) Trouble Ahead [online]. Available from: http//www.education. guardian.co.uk/print/0,3858,4516621-101350,00.html [Accessed 28 November 2003]. London: *Guardian.*

Secretary of State for Education and Science. (1966) *Plan for Polytechnics and other Colleges: Higher Education in the Further Education System.* London: HMSO.

See, L. Openshaw, S. (1998) Some Empirical Experiments in Building Fuzzy Models of Spatial Data. In S. Carver (eds) *Innovations in GIS 5.* London: Taylor Francis.

Shevky, E., Bell, W. (1955) *Social Area Analysis.* California: Standford University Press.

Shevky, E., Williams, M. (1949) *The Social Area of Los Angeles.* Los Angeles: University of California.

Shuttleworth, I. (1995) The Relationship Between Social Deprivation, as Measured by Individual Free School Meal Eligibility, and Educational Attainment at GCSE in Northern Ireland: A Preliminary Investigation. *British Educational Research Journal.* 21(4), 487–504.

Simpson, E.H. (1949) Measurement of Diversity. *Nature.* 163, 688.

Singleton, A. (2003) Where do Manchester University Students come from? A Geodemographic Analysis. Unpublished Dissertation. Manchester: Manchester University.

Singleton, A.D., Farr, M. (2004) Widening Access and Participation in Higher Education. Proceedings of the GIS Research UK 12th Annual Conference. 28–30 April 2004. University of East Anglia, Norwich.

Sleight, P. (1993) *Targeting Customers. How to use Geodemographic and Lifestyle Data in Your Business.* Exeter: NTC.

Smith, D. (2007) Britain's Social Mobility: It's Not as Bad as You Think. 1 July 2007, 9. London: *The Sunday Times.*

Smithers, A. (2002) Privatisation gets Public Study Centre. 5 April 2002. London: *The Times Higher Education Supplement.*

Stafford, M., Marmot, M. (2003) Neighbourhood Deprivation and Health: Does it Affect us all Equally? *International Journal of Epidemiology.* 32(3), 357–366.

Sui, D.Z. (1998) Deconstructing Virtual Cities: From Reality to Hyperreality. *Urban Geography.* 19(7), 657–676.

Sullivan, A. (2001) Cultural Capital and Educational Attainment. *Sociology.* 35(4), 893–912.

Swartz, D. (1997) *Culture and Power, the Sociology of Pierre Bourdieu.* Chicago: University of Chicago Press.

Talbot, R. (1991) Underprivileged Areas and Health Care Planning: Implications of use of Jarman Indicators of Health Care Provision. *British Medical Journal.* 302, 383–386.

The Economist. (2004) Who Pays to Study? 24 January 2003, 23.

Thomas, K. (1990) *Gender and Subject in Higher Education.* London: Taylor Fancis.

Thomas, L. (2002) Student Retention in Higher Education: The Role of Institutional Habitus. *Journal of Educational Policy.* 17 (4), 423–442.

Tickle, M., Moulding, G., Milsom, K., Blinkhorn, A. (2000) Socioeconomic and Geographical Influences on Primary Dental Care Preferences in a Population of Young Children. *British Dental Journal* 188(10), 559–562.

Tonks, D. (1999) Access to UK Higher Education, 1991–98: Using Geodemographics. *Widening Participation and Lifelong Learning.* 17(1), 26–36.

Tonks, D., Farr, M. (1995) Market Segments for Higher Education: Using Geodemographics. *Market Intelligence and Planning.* 13(4), 24–33.

Tooley, J. (1997) On School Choice and Social Class: A Response to Ball, Bowe and Gewirtz. *British Journal of Sociology of Education.* 18 (2), 217–230.

Tough, S., Brooks, R. (2007) *School Admissions: Fair choice for Parents and Pupils.* London: IPPR. June 2007.

Townsend, P., Beatie, A. (1988) *Health and Deprivation. Inequality and the North.* London: Croom Helm.

Tranmer, M., Steel, D.G. (1998) Using Census Data to Investigate the Cause of the Ecological Fallacy. *Environment and Planning A.* 30, 817–831.

UCCA (1994) *Its Origins and Development 1950–93.* London: H.E. Body and Co.

Vanderwal, T. (2007) Folksonomy [online]. Available from: www.vanderwal.net/folksonomy.html [Accessed 21 July 2007].

Vickers, D. (2006) Multi-level Integrated Classifications Based on the 2001 Census. Unpublished PhD Thesis, University of Leeds.

Vickers, D. and Rees, P. (2007) Creating the National Statistics 2001 Output Area Classification. *Journal of the Royal Statistical Society, Series A.* 170(2), 370–403.

Vickers, D., Rees, P., Birkin, M. (2005) Creating the National Classification of Census Output Areas: Data Methods and Results. [online]. The School of Geography, University of Leeds Working Paper. Available from www.geog.leeds.ac.uk/wpapers/05-2.pdf [Accessed 21 August 2006]. Leeds: University of Leeds.

Voas, D., Williamson, P. (2001) The Diversity of Diversity: A Critique of Geodemographic Classification. *Area.* 33(1), 63–76.

Walker, M. (2003) Framing Social Justice in Education: What Does the Capabilities Approach Offer? *British Journal of Educational Studies.* 51(2), 168–187.

Wallace, M., Charlton, J., Denham, C. (1995) The new OPCS area classifications. *Population Trends*. 79, 15–30.

Ward, J.H. (1963) Hierarchical grouping to optimise an objective function. *Journal of the American Statistical Association*. 58, 236–234.

Ward, L. (2003) Test that Discounts Privileged. *The Guardian*. 20 October 2003.

Ward, L., Smithers, R. (2004) Comprehensives Top "Value Added" Tables [online] Available from: www.education.guardian.co.uk/schools/story/0,5500,1123098,00.html [Accessed 1 June 2007]. London: *The Guardian*.

Webber, R. (1977) Technical Paper 23: An Introduction to the National Classification of Wards and Parishes. London: Centre for Environmental Studies.

Webber, R. (1978) Parliamentary Constituencies: a Socio-Economic Classification. OPCS Occasional Paper 13. London: OPCS.

Webber, R. (2004) The Relative Power of Geodemographics Vis-à-vis Person Level Demographic Variables as Discriminators of Consumer Behaviour. CASA Working Paper Series. 84.

Webber, R. (2007) The Metropolitan Habitus: Its Manifestations, Locations, and Consumption Profiles. *Environment and Planning A*. 39(1), 182–207.

Webber, R., Craig, J. (1978) *Studies in Medical and Other Population Subjects, 35: Socio-economic Classifications of Local Authority Areas*. London: OPCS.

Weber, M. (1920) *Economy Society*. California: University of California Press.

Weinberger, D. (2007) *Everything is Miscellaneous: The Power of the New Digital Disorder*. Baltimore: Henry Holt and Company.

Willets, D. (2007) Speech to the CBI on the Conservatives' Backing for Labour's Academy System [online]. Available from: www.news.bbc.co.uk/1/hi/uk_politics/6662219.stm [Accessed 1 June 2007]. London: CBI.

Wilson, A. (1972) Theoretical Geography: Some Speculations. *Transactions of the Institute of British Geographers*. 57, 31–44.

Wrigley, N., Holt, T., Steel, D., Tranmer, M. (1996) Analysing, Modelling and Resolving the Ecological Fallacy. In P. Longley, M. Batty (eds). *Spatial Analysis: Modelling in a GIS Environment*. New York, USA: Wiley.

Zadeh, L. (1965) Fuzzy Sets. *Information and Control*. 8, 338–353.

Index